ANALYTICAL CONICS

Barry Spain

Dover Publications, Inc.
Mineola, New York

Bibliographical Note

This Dover edition, first published in 2007, is an unabridged republication of the work originally published in 1957 by Pergamon Press, London and New York.

International Standard Book Number: 0-486-45773-7

Manufactured in the United States of America
Dover Publications, Inc., 31 East 2nd Street, Mineola, N.Y. 11501

PREFACE

DESPITE the many textbooks written on analytical geometry the novice in this field is unlikely to obtain a *concise* book which begins with the elements and ranges over the basic ideas and methods. My aim, then, is to provide a compact book, readily intelligible to any student with a sound mathematical background. Accordingly this volume is designed both for undergraduates and for pupils who are specializing in mathematics at school. I hope that the conscientious reader will feel stimulated to continue his geometrical studies with the help of the more advanced treatises.

The book starts by developing *ab initio* the theory of the straight line, circle and the conics in their standard forms. A few footnotes are supplied to help the student who is unfamiliar with the elements of determinant theory. Next, the transformation of axes and their invariants are discussed and later used to classify the conic given by the general equation of the second degree. At this stage the line at infinity is introduced with the aid of homogeneous Cartesian coordinates; subsequently, the properties of the conic given by the general equation and pencils of conics are studied. This is followed by a short introduction to cross-ratio and homographic correspondence. Further the concept of line-coordinates and the principle of duality are presented and application of the former is made to the theory of confocal conics. The text ends with a brief exposition of generalized homogeneous coordinates.

There is an Appendix containing worked-out solutions, but the reader is strongly recommended to solve the examples for himself. Only as a last resort should he seek guidance from the Appendix.

My thanks are due to the Board of Trinity College, University of Dublin, for permission to include many examples selected from the University examination papers. For valuable help in reading the proofs I am much indebted to my wife and to Mr. R. E. Harte who has also checked the examples. Finally, I wish to express my appreciation to the staff of Messrs. Pergamon Press for their courtesy and help.

B. S.

Trinity College,
Dublin.
May, 1957

CONTENTS

I. INTRODUCTION

V. HYPERBOLA

VI. PARABOLA

VII. TRANSFORMATION OF AXES

VIII. GENERAL EQUATION OF THE SECOND DEGREE

IX. LINE AT INFINITY

X. CONIC

XI. PENCIL OF CONICS

XII. HOMOGRAPHIC CORRESPONDENCE

XIII. LINE-COORDINATES

XIV. GENERALIZED HOMOGENEOUS COORDINATES

APPENDIX

CHAPTER I

INTRODUCTION

1. Point coordinates

Choose two mutually perpendicular straight lines $X'OX$ and $Y'OY$ (Fig. 1), called the x-axis and y-axis respectively. We make the convention that the directions OX and OY are to be considered as positive, whilst the opposite directions OX' and OY' are to be

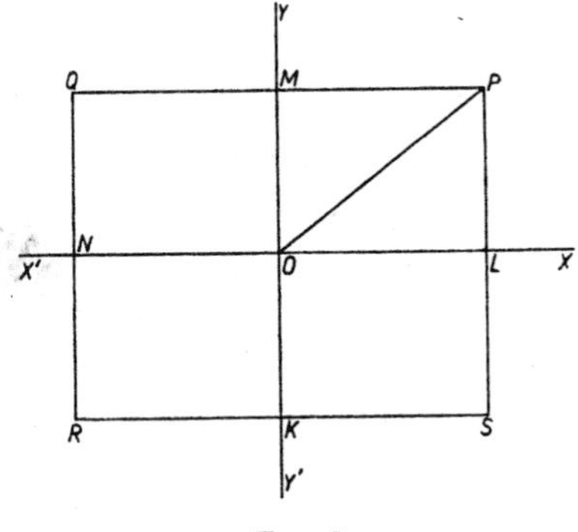

FIG. 1

considered as negative. Then the position of a point P in the plane of the axes is *uniquely* determined by the perpendicular distances MP and LP to the y-axis and x-axis. These distances, which we shall denote by x and y respectively, are called the coordinates of the point P with respect to the given axes. The coordinate MP is sometimes called the **abscissa** of P and LP the **ordinate** of P. We shall write $P \equiv (x, y)$ or $P \equiv (x_P, y_P)$ if we wish to emphasize that x_P and y_P are the coordinates of P.

Note carefully that coordinates are signed quantities, their signs depending on the quadrant in which P lies. For example, draw the rectangle $PMQNRKSL$ in which $QM = MP$ and $SL = LP$. Then $Q \equiv (-x_P, y_P)$, $R \equiv (-x_P, -y_P)$ and $S \equiv (x_P, -y_P)$. Further, we have $L \equiv (x_P, 0)$, $M \equiv (0, y_P)$, $N \equiv (-x_P, 0)$, $K \equiv (0, -y_P)$ and $O \equiv (0, 0)$.

The distance of the point P from the point of intersection O, called the **origin**, of the axes is $OP = \sqrt{x_P^2 + y_P^2}$.

2. Distance between two points

In Fig. 2, draw PL and QM perpendicular to the x-axis and PN parallel to the x-axis. Then PNQ is a right-angled triangle in which $PQ^2 = PN^2 + NQ^2$. But $PN = x_Q - x_P$ and $NQ = y_Q - y_P$. Thus $PQ^2 = (x_Q - x_P)^2 + (y_Q - y_P)^2$. That is,

$$PQ = \sqrt{(x_Q - x_P)^2 + (y_Q - y_P)^2}.$$

Example 1. Calculate the distances between the following pairs of points: (i) (1, 2), (2, 3); (ii) $(-4, 1)$, $(-1, 5)$; (iii) (t^2, t), $(1, 3t)$.

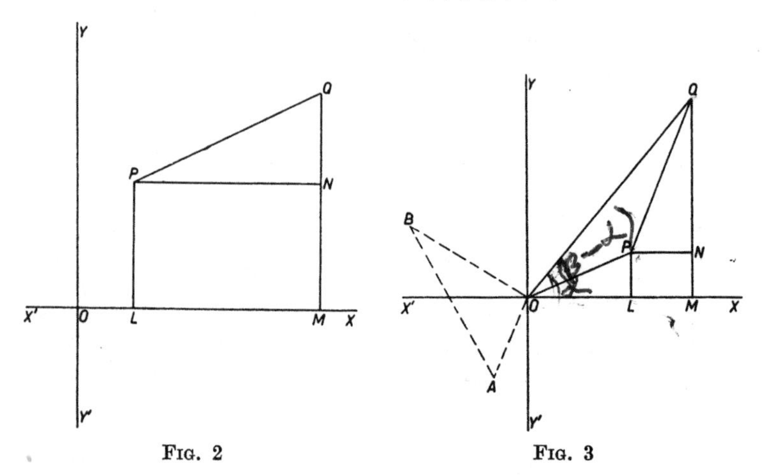

FIG. 2 FIG. 3

3. Area of a triangle

First let us calculate the area of a triangle OPQ (Fig. 3) with one vertex at the origin. Draw PL and QM perpendicular to the x-axis and PN parallel to the x-axis. Then the area of the triangle OPQ is given by

$$\begin{aligned}\Delta OPQ &= \Delta OMQ - \Delta OLP - \Delta PNQ - \text{rectangle } LMNP,\\ &= \tfrac{1}{2}x_Qy_Q - \tfrac{1}{2}x_Py_P - \tfrac{1}{2}(x_Q - x_P)(y_Q - y_P) - y_P(x_Q - x_P),\\ &= \tfrac{1}{2}(x_Py_Q - x_Qy_P). \qquad (3.1)\end{aligned}$$

We note that the interchange of P and Q changes the sign but not the magnitude of the area of the triangle OPQ. Let us introduce the angles $LOP = \alpha$ and $LOQ = \beta$, and we have

$$\begin{aligned}\Delta OPQ &= \tfrac{1}{2}OP \,.\, OQ(\cos\alpha \sin\beta - \cos\beta \sin\alpha),\\ &= \tfrac{1}{2}OP \,.\, OQ \sin(\beta - \alpha).\end{aligned}$$

In order that the area of the triangle OPQ be positive, it is necessary and sufficient that $0 < \beta - \alpha < \pi$. In this case, the cyclic order of the vertices OPQ is counter-clockwise. The dotted triangle OAB is an example of a triangle for which the cyclic order of its vertices is clockwise, and so formula (3.1) will produce a negative area.

Now let us calculate the area of the triangle PQR (Fig. 4). We have

$$\Delta PQR = \Delta OPQ - \Delta ORQ - \Delta OPR.$$

In this formula we have been meticulous to arrange the vertices of

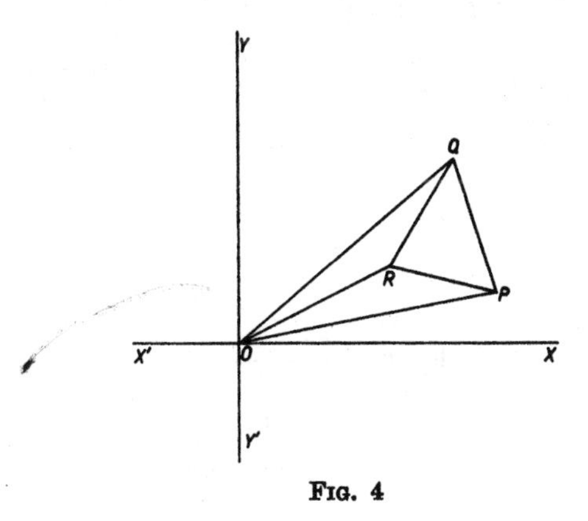

FIG. 4

each triangle in counter-clockwise cyclic order, so that the areas are all positive. On applying formula (3.1), the result is

$$\Delta PQR = \tfrac{1}{2}(x_P y_Q - x_Q y_P + x_Q y_R - x_R y_Q + x_R y_P - x_P y_R).$$

The reader familiar with determinant† theory will see that this equation can be written

$$\Delta PQR = \tfrac{1}{2}\begin{vmatrix} x_P & y_P & 1 \\ x_Q & y_Q & 1 \\ x_R & y_R & 1 \end{vmatrix}. \tag{3.2}$$

† The reader unacquainted with determinants need only note that

$$\begin{vmatrix} x_P & y_P & z_P \\ x_Q & y_Q & z_Q \\ x_R & y_R & z_R \end{vmatrix}$$

is an abbreviation for the expression

$$(x_P y_Q z_R + x_Q y_R z_P + x_R y_P z_Q - x_Q y_P z_R - x_R y_Q z_P - x_P y_R z_Q).$$

The points P, Q and R are collinear, if and only if the area of the triangle PQR is zero. That is, the necessary and sufficient condition that the three points P, Q and R be collinear is

$$\begin{vmatrix} x_P & y_P & 1 \\ x_Q & y_Q & 1 \\ x_R & y_R & 1 \end{vmatrix} = 0. \tag{3.3}$$

Example 2. Calculate the area of the triangles with vertices at (i) $(1, 2)$, $(-2, 1)$, $(-3, -2)$; (ii) $(-2, 3)$, $(4, 5)$, $(1, 1)$.

Example 3. Obtain λ so that the area of the triangle with vertices at $(1, \lambda)$, $(\lambda, 1)$ and (λ, λ) is $+ 12\frac{1}{2}$.

Example 4. Show that the three *distinct* points (a^2, a), (b^2, b) and (c^2, c) can never be collinear.

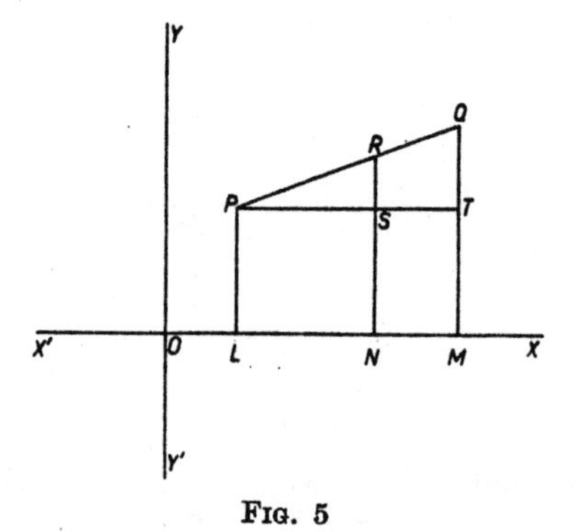

FIG. 5

4. Joachimsthal's section formula

We often require the coordinates of the point R which divides the line joining P and Q in the ratio $PR:RQ = \lambda:\mu$. In Fig. 5, draw PL, RN and QM perpendicular to and PST parallel to the x-axis. From the similar triangles PSR and PTQ we have

$$PS:PT = PR:PQ = \lambda:(\lambda + \mu).$$

That is,

$$(x_R - x_P):(x_Q - x_P) = \lambda:(\lambda + \mu).$$

Likewise, we obtain

$$(y_R - y_P):(y_Q - y_P) = \lambda:(\lambda + \mu).$$

Solving these equations for x_R and y_R, we deduce Joachimsthal's section formulae

$$x_R = \frac{\mu x_P + \lambda x_Q}{\mu + \lambda}; \quad y_R = \frac{\mu y_P + \lambda y_Q}{\mu + \lambda}. \tag{4.1}$$

These formulae also apply when R lies outside PQ, in which case the ratio $\lambda:\mu$ is negative, being numerically greater or less than unity according as Q lies between P and R or P lies between R and Q respectively. The ratio $\lambda:\mu$ never equals -1 for any finite position of R.

Example 5. Show that the coordinates of the mid-point of PQ are $(\frac{1}{2}(x_P + x_Q), \frac{1}{2}(y_P + y_Q))$.

Example 6. Prove that the medians of the triangle PQR are concurrent at the point $(\frac{1}{3}(x_P + x_Q + x_R), \frac{1}{3}(y_P + y_Q + y_R))$.

Example 7. In what ratio does the point $(-1, -1)$ divide the join of $(-5, -3)$ and $(5, 2)$.

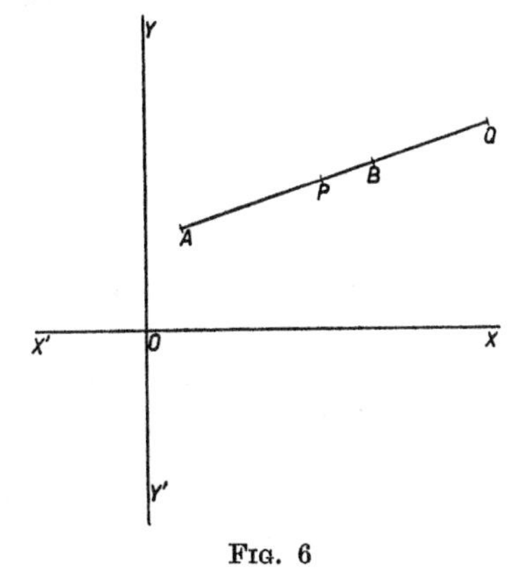

FIG. 6

5. Harmonic range of points

If four collinear points A, B, P and Q (Fig. 6) are such that

$$AP:PB = -AQ:QB,$$

then P and Q are called **harmonic conjugates** with respect to A and B, and A, B, P and Q are said to form a **harmonic range.** We can rewrite the above ratio equation in the form

$$PB:BQ = -PA:AQ,$$

which shows that A and B are also harmonic conjugates with respect to P and Q.

If $AP:PB = \lambda:\mu$ then by Joachimsthal's section formula (4.1) we have

$$P \equiv \left(\frac{\lambda x_B + \mu x_A}{\lambda + \mu}, \frac{\lambda y_B + \mu y_A}{\lambda + \mu}\right), \quad Q \equiv \left(\frac{\lambda x_B - \mu x_A}{\lambda - \mu}, \frac{\lambda y_B - \mu y_A}{\lambda - \mu}\right).$$

Note that there is no finite point conjugate to the mid-point of AB.

Example 8. Given $A \equiv (-3, -2)$, $B \equiv (4, 12)$, calculate the coordinates of P if $AP:PB = 3:4$. Further, obtain the coordinates of Q, where P and Q are harmonic conjugates to A and B.

Example 9. Prove that the two pairs of points $(\alpha, 0)$, $(\beta, 0)$ and $(\gamma, 0)$, $(\delta, 0)$ are harmonic conjugates of each other, where α and β are the roots of the quadratic equation $ax^2 + 2hx + b = 0$ and γ, δ are the roots of $a'x^2 + 2h'x + b' = 0$ if $ab' + a'b - 2hh' = 0$.

Answers

1. (i) $\sqrt{2}$, (ii) 5, (iii) $t^2 + 1$. **2.** (i) 4, (ii) -9. **3.** -4 or 6. **7.** 2:3. **8.** (0, 4); $(-24, -44)$.

CHAPTER II

STRAIGHT LINE

6. Straight line

If $P \equiv (x, y)$ is any point on the straight line AB, then P, A and B are collinear, and so by (3.3) we have

$$\begin{vmatrix} x & y & 1 \\ x_A & y_A & 1 \\ x_B & y_B & 1 \end{vmatrix} = 0. \tag{6.1}$$

This equation is of the form

$$lx + my + n = 0, \tag{6.2}$$

which is a linear equation in x and y.

Conversely, what is the locus of a point P whose coordinates x and y satisfy the linear equation (6.2)? Select three points, A, B and C on the locus and we have

$$lx_A + my_A + n = 0, \tag{6.3}$$

$$lx_B + my_B + n = 0, \tag{6.4}$$

$$lx_C + my_C + n = 0. \tag{6.5}$$

The elimination† of l, m and n yields the determinantal result

$$\begin{vmatrix} x_A & y_A & 1 \\ x_B & y_B & 1 \\ x_C & y_C & 1 \end{vmatrix} = 0.$$

That is, the three points A, B and C are collinear. In the same way, all possible triples of points on the locus (6.2) are collinear. Therefore, the locus represented by equation (6.2) is a straight line.

We have thus established the important result that **every linear equation represents a straight line.**

Let ψ be the angle which AB makes with the positive direction

† To eliminate, solve (6.3) and (6.4) for the ratios

$$l:m:n = (y_A - y_B):(x_B - x_A):(x_A y_B - x_B y_A)$$

and substitute in (6.5).

of the x-axis (Fig. 7). The convention chosen is that positive angles correspond to counter-clockwise rotation from the positive x-axis to the straight line AB. (The angle which BA makes with the x-axis is then $\pi + \psi$.) The tangent of this angle ψ is called the **gradient** or **slope** of the straight line AB. Draw AL and BL parallel to the x-axis and y-axis respectively. Then we see that

$$g = \tan \psi = \frac{LB}{AL} = \frac{y_B - y_A}{x_B - x_A}.$$

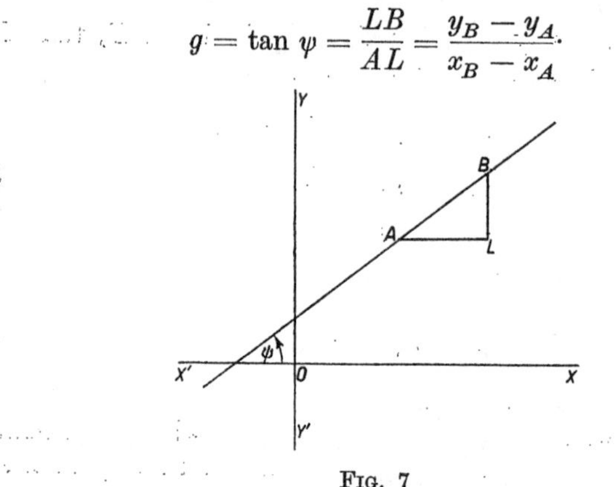

FIG. 7

Now subtract equation (6.3) from (6.4) and the result is

$$l(x_B - x_A) + m(y_B - y_A) = 0$$

and so

$$g = \tan \psi = -\frac{l}{m}.$$

That is, the gradient of the straight line $lx + my + n = 0$ is $-l/m$.

Example 1. **The equation of the straight line through A with gradient g is $y - y_A = g(x - x_A)$.**

Example 2. **The equation of the straight line of gradient g which intercepts a distance c on the y-axis is $y = gx + c$.**

Example 3. **The equation of the straight line which intercepts distances a and b on the x-axis and y-axis respectively is $\frac{x}{a} + \frac{y}{b} = 1$.**

Example 4. **The equation of the straight line through A and B can be written in the form $\frac{x - x_A}{y - y_A} = \frac{x_B - x_A}{y_B - y_A}$.**

7. Parametric form

Consider the two equations

$$x = x_A + pt; \; y = y_A + qt, \tag{7.1}$$

where p, q, x_A and y_A are constants but t is a variable. The elimination of t yields the linear equation

$$q(x - x_A) = p(y - y_A)$$

and so (7.1) represents a straight line through A with gradient q/p. The equations (7.1) are called the **parametric equations** of the straight line and t is called the **parameter.**

Example 5. Show that the distance between the points $P \equiv (x, y)$ and A is given by $\sqrt{p^2 + q^2}\, t$.

Example 6. The parametric equations of the straight line through A with gradient $\tan \psi$ can be written in the form

$$x = x_A + t \cos \psi; \quad y = y_A + t \sin \psi,$$

where t now measures the distance between $P \equiv (x, y)$ and A.

Example 7. Prove that $x = x_A + t(x_B - x_A)$; $y = y_A + t(y_B - y_A)$ are parametric equations of the straight line AB. What is the geometrical significance of t?

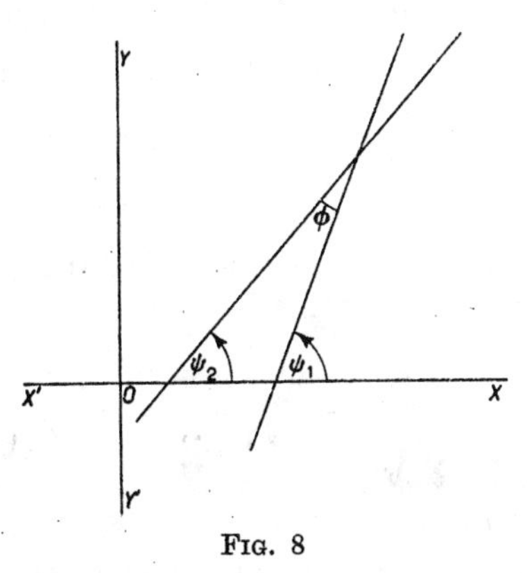

FIG. 8

8. Angle between two straight lines

We wish to calculate the angle φ between the two straight lines

$$l_1 x + m_1 y + n_1 = 0, \tag{8.1}$$

$$l_2 x + m_2 y + n_2 = 0, \tag{8.2}$$

which make angles ψ_1 and ψ_2 respectively with the x-axis (Fig. 8). We have

$$\tan \psi_1 = -\frac{l_1}{m_1}; \quad \tan \psi_2 = -\frac{l_2}{m_2}; \quad \varphi = \psi_1 - \psi_2.$$

Therefore,

$$\tan \varphi = \tan (\psi_1 - \psi_2) = \frac{\tan \psi_1 - \tan \psi_2}{1 + \tan \psi_1 \tan \psi_2} = \frac{-\dfrac{l_1}{m_1} + \dfrac{l_2}{m_2}}{1 + \dfrac{l_1 l_2}{m_1 m_2}}.$$

That is,

$$\tan \varphi = \frac{l_2 m_1 - l_1 m_2}{l_1 l_2 + m_1 m_2}.$$

This formula yields the acute or obtuse angle between the two straight lines according as ψ_1 is greater than or less than ψ_2 respectively.

If the two straight lines are parallel $\varphi = 0$, and so the necessary and sufficient condition that the two straight lines (8.1) and (8.2) are parallel is $l_2 m_1 - l_1 m_2 = 0$ or $\dfrac{l_1}{m_1} = \dfrac{l_2}{m_2}$. This result also follows from the fact that parallel straight lines have the same gradient.

Further, the two straight lines are perpendicular to one another if $\varphi = \dfrac{\pi}{2}$ and so the necessary and sufficient condition that the two straight lines (8.1) and (8.2) be mutually perpendicular is $l_1 l_2 + m_1 m_2 = 0$. This condition can also be written $\left(-\dfrac{l_1}{m_1}\right)\left(-\dfrac{l_2}{m_2}\right) = -1$, which states that **the product of the gradients of two perpendicular straight lines is -1.**

Example 8. **The straight line through A perpendicular to the straight line $lx + my + n = 0$ is represented by $m(x - x_A) - l(y - y_A) = 0$.**

Example 9. **Show that the angle between the two straight lines $y = g_1 x + c_1$ and $y = g_2 x + c_2$ is $\tan^{-1} \{(g_1 - g_2)/(1 + g_1 g_2)\}$.**

Example 10. Prove that the four straight lines $4x - 3y - 5 = 0$, $7x + y - 40 = 0$, $x - 2y - 10 = 0$ and $x + 3y + 10 = 0$ form the sides of a cyclic quadrilateral.

9. Intersection of two straight lines

The point P of intersection of the two straight lines (8.1) and (8.2) has co-ordinates x and y which satisfy both equations. Hence the solution of the two simultaneous equations yields the point of intersection at $\left(\dfrac{m_1 n_2 - m_2 n_1}{l_1 m_2 - l_2 m_1}, \dfrac{n_1 l_2 - n_2 l_1}{l_1 m_2 - l_2 m_1}\right)$. There is a unique finite point of intersection of the two straight lines unless the denominator $l_1 m_2 - l_2 m_1$ is zero, in which case the straight lines are parallel.

Example 11. Obtain the coordinates of the vertices of the triangle formed by the three straight lines $2x + 3y - 1 = 0$, $x - 5y - 7 = 0$ and $3x - 2y + 5 = 0$. Hence calculate the area of this triangle.

10. Concurrency

The three straight lines

$$l_1x + m_1y + n_1 = 0, \tag{10.1}$$

$$l_2x + m_2y + n_2 = 0, \tag{10.2}$$

$$l_3x + m_3y + n_3 = 0, \tag{10.3}$$

will be concurrent at the point $P \equiv (x, y)$ if x and y satisfy these three equations. The elimination of x and y (compare §6) yields the necessary condition for concurrency that

$$\begin{vmatrix} l_1 & m_1 & n_1 \\ l_2 & m_2 & n_2 \\ l_3 & m_3 & n_3 \end{vmatrix} = 0. \tag{10.4}$$

Conversely, suppose equation (10.4) is satisfied. If $l_1m_2 - l_2m_1 \neq 0$, the solution of equations (10.1) and (10.2) is $x = \dfrac{m_1n_2 - m_2n_1}{l_1m_2 - l_2m_1}$, $y = \dfrac{n_1l_2 - n_2l_1}{l_1m_2 - l_2m_1}$. By virtue of (10.4) it is easily seen that this point (x, y) also lies on the line (10.3). Thus the three straight lines are concurrent. If, on the other hand, $l_1m_2 - l_2m_1 = 0$ we have $l_1 = l_2m_1/m_2$, and (10.4) reduces to $n_1(l_2m_3 - l_3m_2) + n_2(l_3m_1 - l_1m_3) = 0$. Substitute for l_1 and the result is $(l_2m_3 - l_3m_2)\left(n_1 - \dfrac{n_2m_1}{m_2}\right) = 0$. Hence either (*a*) $l_1m_2 - l_2m_1 = 0$ or (*b*) $n_1 - \dfrac{n_2m_1}{m_2} = 0$. If (*a*) is true, then $\dfrac{l_1}{m_1} = \dfrac{l_2}{m_2} = \dfrac{l_3}{m_3}$. That is, the three straight lines are all parallel to one another. On the other hand, if (*b*) is true, we have $\dfrac{l_1}{l_2} = \dfrac{m_1}{m_2} = \dfrac{n_1}{n_2}$. That is, the straight lines (10.1) and (10.2) are coincident. This proof needs further investigation if $m_2 = 0$ and $l_1m_2 - l_2m_1 = 0$. Since l_2 and m_2 cannot both be zero we have $m_1 = 0$ and (10.4) reduces to $m_3(n_1l_2 - n_2l_1) = 0$. Either $m_3 = 0$ and again the three straight lines are parallel or $n_1l_2 - n_2l_1 = 0$ and the straight lines (10.1) and (10.2) are coincident.

Thus equation (10.4) is a *necessary* but not a sufficient condition for the concurrency of the three given straight lines.

Example 12. Verify that the straight lines $x + 2y = 1$, $3x - 2y = -4$ and $x - 6y = -6$ are concurrent.

Example 13. Show that a necessary condition for the concurrency of the three straight lines $ax + by + c = 0$, $bx + cy + a = 0$ and $cx + ay + b = 0$ is $a^3 + b^3 + c^3 = 3abc$.

11. Pencil of straight lines

Consider the two distinct straight lines given by

$$u_1 \equiv l_1x + m_1y + n_1 = 0,$$
$$u_2 \equiv l_2x + m_2y + n_2 = 0,$$

and set up the equation

$$\lambda u_1 + \mu u_2 = 0. \qquad (11.1)$$

This equation is linear in x and y and so represents a straight line. Further, the coordinates of the point of intersection of $u_1 = 0$ and $u_2 = 0$, obtained by solving the simultaneous equations $u_1 = u_2 = 0$, also satisfy equation (11.1). Thus (11.1) represents a straight line passing through the point of intersection of $u_1 = 0$ and $u_2 = 0$. As the ratio $\lambda:\mu$ varies, equation (11.1) represents all the straight lines through this point of intersection. We say that these straight lines form a **pencil** of straight lines and sometimes refer to $u_1 = 0$ and $u_2 = 0$ as the **base** lines of the pencil. Clearly any two straight lines of the pencil can be chosen as base lines.

The straight line $u_3 \equiv l_3x + m_3y + n_3 = 0$ is a member of the pencil if†

$$\lambda u_1 + \mu u_2 + \nu u_3 = 0 \qquad (11.2)$$

and so a necessary condition that the three straight lines be concurrent is that multiples λ, μ and ν exist such that equation (11.2) is satisfied.

If $u_1 = 0$ and $u_2 = 0$ are parallel, then $\lambda u_1 + \mu u_2 = 0$ represents (as $\lambda:\mu$ varies) all the straight lines parallel to $u_1 = 0$ or $u_2 = 0$. Hence (11.2) is not a sufficient condition for concurrency. The three straight lines might be parallel to one another.

Example 14. Obtain the equation of the straight line through the origin and through the point of intersection of the straight lines $x + y - 3 = 0$ and $2x = y + 5$.

Example 15. Show that the three altitudes of a triangle are concurrent at a point called the **orthocentre.** (Hint: the altitudes of the triangle ABC are $u_1 \equiv (x - x_A)(x_B - x_C) + (y - y_A)(y_B - y_C) = 0$, etc., and $u_1 + u_2 + u_3 = 0$.)

Example 16. Show that the three medians of a triangle are concurrent at a point called the **centroid.**

† Note with care that if the equations $ax + by + c = 0$ and $px + qy + r = 0$ represent the same straight line, then $a:b:c = p:q:r$.

Example 17. Prove that the three perpendicular bisectors of the sides of a triangle are concurrent at a point called the **circumcentre.** (The circumcentre is so called because it is the centre of the circle circumscribed about the triangle.)

Example 18. Obtain the coordinates of (i) the centroid, (ii) the orthocentre and (iii) the circumcentre of the triangle formed by the three straight lines $2x + y - 31 = 0$, $3x - y - 4 = 0$ and $31x - 17y - 188 = 0$. Show that the centroid divides the join of the orthocentre and circumcentre in the ratio 2:1.

12. Sign of the expression $u \equiv lx + my + n$

Let the straight line AB cut the straight line $u = 0$ at P, where $AP:PB = \lambda:\mu$. Then by Joachimsthal's section formula (4.1)

$$P \equiv \left(\frac{\mu x_A + \lambda x_B}{\mu + \lambda}, \frac{\mu y_A + \lambda y_B}{\mu + \lambda}\right).$$

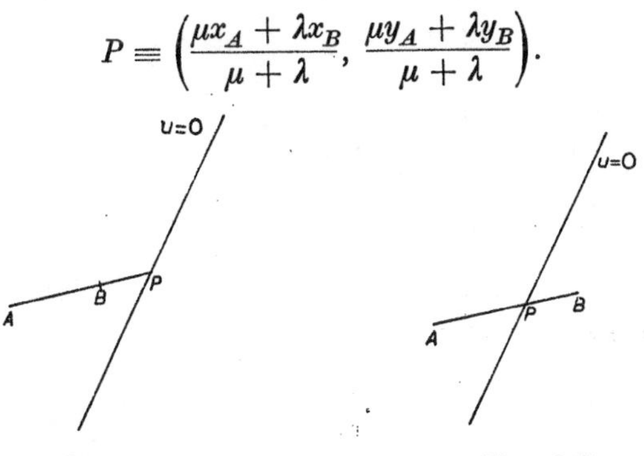

FIG. 9 (a)

FIG. 9 (b)

Since P lies on $u = 0$, we have

$$\frac{l(\mu x_A + \lambda x_B)}{\mu + \lambda} + \frac{m(\mu y_A + \lambda y_B)}{\mu + \lambda} + n = 0,$$

from which we obtain† that

$$\frac{\lambda}{\mu} = -\frac{lx_A + my_A + n}{lx_B + my_B + n} = -\frac{u_A}{u_B} \text{ (say)}.$$

If u_A and u_B have the same sign, the ratio $\lambda:\mu$ is negative and so P lies outside AB (Fig. 9a). Hence A and B lie on the same side of the straight line $u = 0$. If, on the other hand, u_A and u_B have opposite signs, the ratio $\lambda:\mu$ is positive and so P lies between A and B. Hence A and B lie on opposite sides of $u = 0$ (Fig. 9b).

Thus the straight line $u = 0$ divides the plane into two regions, such that $u > 0$ in one of them and $u < 0$ in the other.

† Note carefully the notation $u_A \equiv lx_A + my_A + n$ but $u_1 \equiv l_1x + m_1y + n_1$.

Example 19. Show that the four points $(4, 1), (1, 2), (-1, -1)$ and $(-2, -3)$ each lie in one of the four regions into which the straight lines $2x - 3y - 4 = 0$ and $x + 2y + 2 = 0$ divide the plane.

13. Perpendicular distance of a point from a straight line

We wish to calculate the perpendicular distance $d = PN$ (Fig. 10) from P to the straight line $u \equiv lx + my + n = 0$. The equation of the straight line PN is $m(x - x_P) - l(y - y_P) = 0$. Since N lies on this line, we have

$$m(x_N - x_P) - l(y_N - y_P) = 0. \qquad (13.1)$$

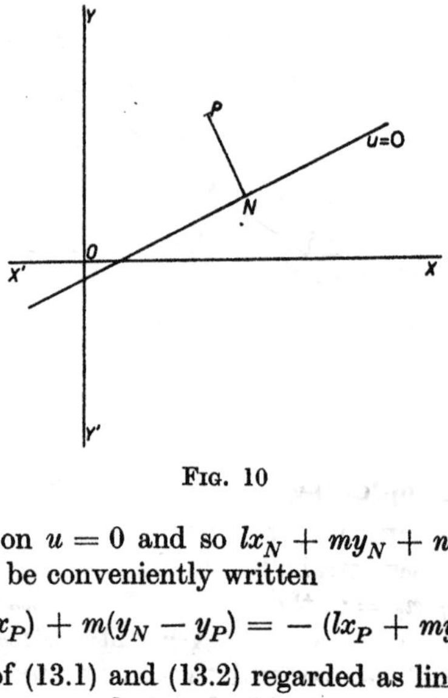

FIG. 10

Further, N lies on $u = 0$ and so $lx_N + my_N + n = 0$, which can for our purposes be conveniently written

$$l(x_N - x_P) + m(y_N - y_P) = -(lx_P + my_P + n). \qquad (13.2)$$

The solution of (13.1) and (13.2) regarded as linear simultaneous equations in $x_N - x_P$ and $y_N - y_P$ is

$$x_N - x_P = -\frac{l(lx_P + my_P + n)}{l^2 + m^2}; \quad y_N - y_P = -\frac{m(lx_P + my_P + n)}{l^2 + m^2}.$$

Thus

$$d^2 = (x_N - x_P)^2 + (y_N - y_P)^2 = \frac{(lx_P + my_P + n)^2}{l^2 + m^2},$$

Hence,

$$d = \pm \frac{lx_P + my_P + n}{\sqrt{l^2 + m^2}}.$$

The sign of d is indeterminate because $u = 0$ and $-u = 0$ represent the same straight line. In the previous section it was shown that the sign of the expression $lx_P + my_P + n$ depends on which side

P is of the line $u = 0$. We can, therefore, make the convention that the sign be taken in such a way that the perpendicular from the origin is positive. That is, d has the same sign as n. This convention, of course, does not apply to straight lines through the origin. For these lines we make, if necessary, any suitable convention.

Example 20. Calculate the perpendicular distances of the point $(1, -2)$ from the straight lines (i) $3x - 4y = 2$, (ii) $5x + 12y = 6$ and (iii) $7x + 8y = 1$.

Example 21. Calculate the altitudes of the triangle in example 18.

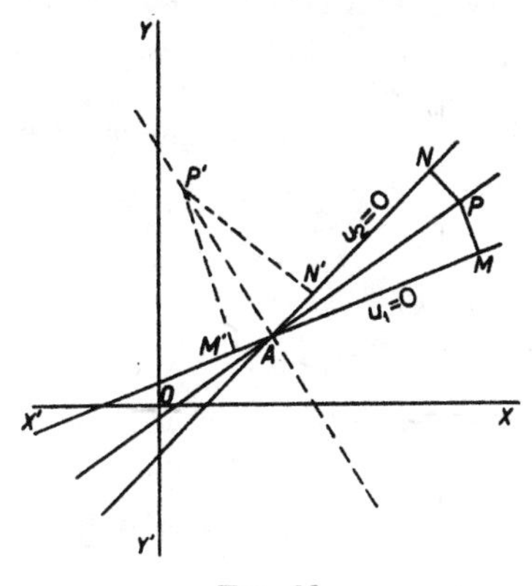

FIG. 11

14. Bisectors of angles between two straight lines

If $P \equiv (x, y)$ is any point on the internal bisector (Fig. 11) of the angle between the two straight lines $u_1 \equiv l_1x + m_1y + n_1 = 0$ and $u_2 \equiv l_2x + m_2y + n_2 = 0$, the perpendicular distances PM, PN to the straight lines are of equal length arithmetically. Likewise if P' is any point on the external bisector (dotted in Fig. 11), the perpendicular distances $P'M'$ and $P'N'$ are equal arithmetically. Thus the pair of equations

$$\frac{l_1x + m_1y + n_1}{\sqrt{l_1^2 + m_1^2}} = \pm \frac{l_2x + m_2y + n_2}{\sqrt{l_2^2 + m_2^2}} \tag{14.1}$$

represent the internal and external bisectors.

If n_1n_2 is positive, then the perpendiculars to $u_1 = 0$ and $u_2 = 0$ from the origin have the same sign. Hence the positive sign in (14.1) corresponds to that bisector of the angle in which the origin lies.

In a numerical example, it is advisable to draw a rough diagram, noting the approximate gradients of the internal and external bisectors. The accurate values obtained from (14.1) are clearly negative reciprocals of one another, since the bisectors are mutually orthogonal.

Example 22. Show that the negative sign in (14.1) corresponds to the bisector of the angle in which the origin lies if $n_1 n_2 < 0$.

Example 23. Find the equation of the internal bisector of the angle between the following pairs of lines: (i) $x + 2y = 1$ and $2x + y + 3 = 0$, (ii) $3x - 4y = 5$ and $-5x + 12y = 2$.

15. Condition that $ax^2 + 2hxy + by^2 + 2gx + 2fy + c = 0$ represents two straight lines

We can represent the two straight lines $u_1 \equiv l_1 x + m_1 y + n_1 = 0$ and $u_2 \equiv l_2 x + m_2 y + n_2 = 0$ by the quadratic equation

$$u_1 u_2 \equiv (l_1 x + m_1 y + n_1)(l_2 x + m_2 y + n_2) = 0.$$

That is,

$$l_1 l_2 x^2 + (l_1 m_2 + l_2 m_1)xy + m_1 m_2 y^2 + (l_1 n_1 + l_2 n_1)x + (m_1 n_2 + m_2 n_1)y + n_1 n_2 = 0.$$

This is a quadratic equation of the second degree and suggests the following question: "Does the general quadratic equation

$$S \equiv ax^2 + 2hxy + by^2 + 2gx + 2fy + c = 0$$

represent two straight lines?"

It is a straightforward calculation to verify the identity

$$(ab - h^2)S \equiv b(ax + hy + g)^2 - 2h(ax + hy + g)(hx + by + f) + a(hx + by + f)^2 + \Delta,$$

where

$$\Delta \equiv abc + 2fgh - af^2 - bg^2 - ch^2. \tag{15.1}$$

If $ab - h^2 \neq 0$, it is clear that S factorizes if and only if $\Delta = 0$. However, if $ab - h^2 = 0$, both a and b cannot vanish as this would imply $h = 0$ and hence S is not a quadratic. Let us first suppose $a \neq 0$, $b \neq 0$, then

$$aS \equiv (ax + hy + g)^2 - 2(af - gh)y + ac - g^2 = 0 \tag{15.2}$$

and it is clear that S now factorizes if and only if $af - gh = 0$. We have

$$\Delta = c(ab - h^2) + f(gh - af) + g(fh - bg)$$

and eliminating b from the last factor by means of $ab - h^2 = 0$, we obtain

$$\Delta = -\frac{1}{a}(gh - af)^2.$$

Thus if S factorizes $\Delta = 0$ and conversely if $\Delta = ab - h^2 = 0$, then $gh - af = 0$ and S factorizes.

It remains to examine the case $b = h = 0$ (or $a = h = 0$). Now we have

$$S \equiv ax^2 + 2gx + 2fy + c = 0$$

and clearly S factorizes if and only if $f = 0$. Thus $\Delta = 0$. Conversely, if $\Delta = b = h = 0$, then $af^2 = 0$. This implies $f = 0$, since $a \neq 0$. Consequently S factorizes.

Hence we have shown that the vanishing of Δ given by (15.1) is the necessary and sufficient condition that $S = 0$ represent two straight lines. This condition can be expressed determinantally

$$\Delta \equiv \begin{vmatrix} a & h & g \\ h & b & f \\ g & f & c \end{vmatrix} = 0. \tag{15.3}$$

When $S = 0$ represents two straight lines, we may write

$$S \equiv \rho(l_1x + m_1y + n_1)(l_2x + m_2y + n_2),$$

where ρ is a factor of proportionality. The straight lines through the origin parallel to $S = 0$ are therefore given by

$$\rho(l_1x + m_1y)(l_2x + m_2y) = 0.$$

That is, the equation of the straight lines through the origin parallel to $S = 0$ is given by

$$ax^2 + 2hxy + by^2 = 0.$$

Note carefully that this equation represents straight lines even if $S = 0$ does not do so. If the coefficients a, b, c, f, g, h are real these lines are real, coincident or conjugate complex according as $ab - h^2$ is negative, zero or positive respectively. We often refer to the straight lines $S = 0$ as a line-pair.

Example 24. For what values of λ do the following equations represent straight lines: (i) $x^2 - 4xy - y^2 + 6x + 8y + \lambda = 0$, (ii) $2\lambda xy - y^2 + 4x + 2y + 8 = 0$.

Example 25. Show that the equation $(x_A{}^2 + y_A{}^2 - r^2)(x^2 + y^2 - r^2) - (x_Ax + y_Ay - r^2)^2 = 0$ represents two straight lines.

16. Angle between line-pair S=0

The angle between the line-pair given by $S = 0$ is the same as that between the line-pair $ax^2 + 2hxy + by^2 = 0$. Let us write

$$ax^2 + 2hxy + by^2 \equiv \rho(l_1x + m_1y)(l_2x + m_2y).$$

Comparing coefficients, we have

$$\rho l_1 l_2 = a;\ \rho(l_1 m_2 + l_2 m_1) = 2h;\ \rho m_1 m_2 = b,$$

from which it follows that

$$\rho^2(l_1 m_2 - l_2 m_1)^2 = \rho^2(l_1 m_2 + l_2 m_1)^2 - 4\rho^2 l_1 l_2 m_1 m_2 = 4(h^2 - ab).$$

The angle between $l_1 x + m_1 y = 0$ and $l_2 x + m_2 y = 0$ is φ, where

$$\tan\varphi = \frac{l_2 m_1 - l_1 m_2}{l_1 l_2 + m_1 m_2}.$$

Hence, by substitution we have

$$\tan\varphi = \pm\frac{2\sqrt{h^2 - ab}}{a + b}. \tag{16.1}$$

The indeterminacy of sign is inherent in this formula because the angle of intersection is φ or $\pi - \varphi$.

We deduce that a line-pair consists of:

(i) two parallel lines if $ab - h^2 = 0$,
(ii) two mutually perpendicular lines if $a + b = 0$.

Example 26. Show that $x^2 + xy - 6y^2 - x - 8y - 2 = 0$ represents a line-pair and calculate their angle of intersection.

17. Equation of lines joining origin to the points of intersection of a line-pair and a straight line

Consider the line-pair

$$S \equiv ax^2 + 2hxy + by^2 + 2gx + 2fy + c = 0$$

and the straight line

$$u \equiv lx + my + n = 0.$$

The coordinates of their points of intersection A and B (Fig. 12) satisfy the equations $S = u = 0$. Now set up the homogeneous equation

$$n^2(ax^2 + 2hxy + by^2) - 2n(gx + fy)(lx + my) + c(lx + my)^2 = 0, \tag{17.1}$$

which is homogeneous of the second degree and so represents two straight lines through the origin. This line-pair is cut by $u = 0$ at the points given by $S = u = 0$. But these are the points A and B of intersection of $S = 0$ and $u = 0$. Hence (17.1) is the required equation of the straight lines joining the origin to the points A and B. Note carefully that this equation is obtained by making $S = 0$ homogeneous by means of $u = 0$.

The result is still valid if $S = 0$ does not represent straight lines. In that event, equation (17.1) represents the equation of the line-pair joining the origin to the two points of intersection of the curve $S = 0$ and the straight line $u = 0$.

Example 27. **Show that the lines joining the origin to the points of intersection of the line-pair of Example 26 and the straight line $x - 6y - 2 = 0$ are mutually perpendicular.**

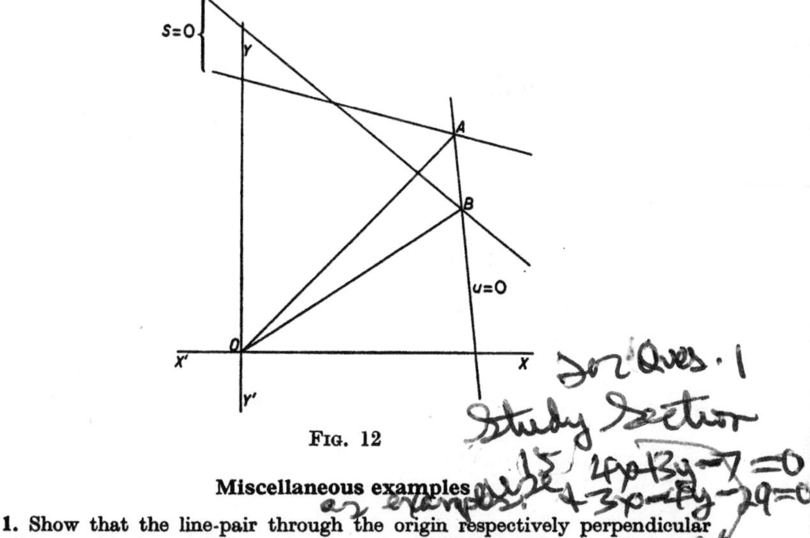

FIG. 12

Miscellaneous examples

1. Show that the line-pair through the origin respectively perpendicular to the line-pair $ax^2 + 2hxy + by^2 = 0$ is given by $bx^2 - 2hxy + ay^2 = 0$.

2. Obtain the coordinates of the mirror image of the point (α, β) in the straight line $lx + my + n = 0$.

3. Prove that the area of the triangle formed by the straight lines $ax^2 + 2hxy + by^2 = 0$ and $lx + my + n = 0$ is $n^2\sqrt{h^2 - ab}/(am^2 - 2hlm + bm^2)$.

4. Show that the equation of the bisectors of the angles between the line-pair $ax^2 + 2hxy + by^2 = 0$ is $h(x^2 - y^2) - (a - b)xy = 0$.

5. Prove that the lines joining the origin to the points of intersection of $ax^2 + 2hxy + by^2 + 2gx + 2fy + c = 0$ and $lx + my + n = 0$ are mutually perpendicular if $n^2(a + b) - 2n(gl + fm) + c(l^2 + m^2) = 0$.

6. Obtain the coordinates of the centre of the circle inscribed in the triangle whose vertices are at $(-9, -8)$, $(15, -2)$ and $(-1, 11)$. Further, calculate the radius of this inscribed circle.

7. Show that the point of intersection of the line-pair $ax^2 + 2hxy + by^2 + 2gx + 2fy + c = 0$ if $ab - h^2 \neq 0$ is always at the real point $((hf - bg)/(ab - h^2), (hg - af)/(ab - h^2))$.

8. The straight line $lx + my + n = 0$ cuts the distinct pair of straight lines $ax^2 + 2\,hxy + by^2 = 0$ at the points P and Q. If the angle OPQ equals the angle OQP, O being the origin, show that $h(l^2 - m^2) = lm\,(a - b)$.

Answers

7. AP/AB. **11.** $(-1, 1)$, $(2, -1)$, $-3, -2)$; $6\frac{1}{2}$. **14.** $x - 8y = 0$. **18.** (i) $(4, 1\frac{1}{3})$, (ii) $(38, 0)$, (iii) $(-13, 2)$. **20.** (i) $-9/5$, (ii) $25/13$, (iii) $10/\sqrt{113}$. **21.** $13\sqrt{5}$, $-2\sqrt{10}$, $26\sqrt{2}/5$. **23.** (i) $3x + 3y + 2 = 0$, (ii) $64x - 112y - 55 = 0$. **24.** (i) -11, (ii) 1 or $-\frac{1}{2}$. **26.** $45°$.

Miscellaneous examples. **2.** $([(m^2 - l^2)\alpha - 2lm\beta - 2nl]/(l^2 + m^2), [-2lm\alpha + (l^2 - m^2)\beta - 2mn]/(l^2 + m^2))$. **6.** $(3/2, 1)$, $3\sqrt{17}/2$.

CHAPTER III

CIRCLE

18. Equation of circle

If $P \equiv (x, y)$ is any point on the circle with centre at C and radius r, we have

$$(x - x_C)^2 + (y - y_C)^2 = r^2, \qquad (18.1)$$

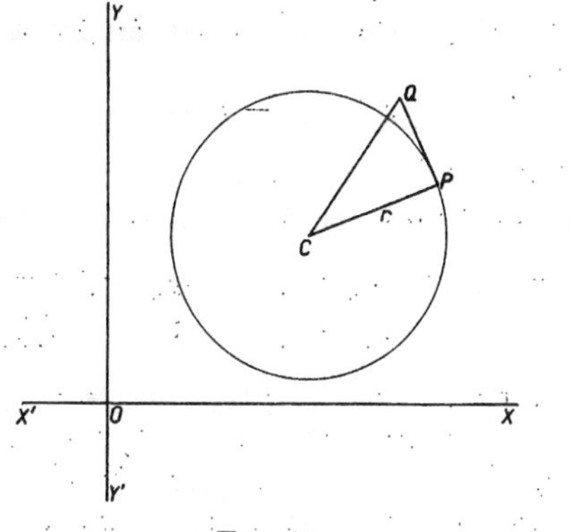

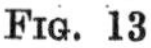

FIG. 13

which can be written

$$x^2 + y^2 - 2x_C x - 2y_C y + x_C{}^2 + y_C{}^2 - r^2 = 0.$$

We see that:

(1) this equation is quadratic,
(2) the coefficient of x^2 is equal to the coefficient of y^2,
(3) there is no term in the product xy.

Conversely, the most general equation satisfying these three conditions is

$$ax^2 + ay^2 + 2gx + 2fy + c = 0. \quad (a \neq 0) \qquad (18.2)$$

We may write this equation in the form

$$\left(x + \frac{g}{a}\right)^2 + \left(y + \frac{f}{a}\right)^2 = \frac{g^2 + f^2 - ac}{a^2},$$

which shows by comparison with (18.1) that it represents a circle with centre at the point $(-g/a, -f/a)$ and radius $\sqrt{g^2 + f^2 - ac}/a$.

The centre of this circle is always a real point (a, f, g and c being real), but its radius is real if and only if $g^2 + f^2 > ac$. If $g^2 + f^2 = ac$, then there is only one real point $(-g/a, -f/a)$ on the circle, and we refer to it as a **point-circle.** If $g^2 + f^2 < ac$ there are no real points on the circle (18.2) and we call it a **virtual** circle.

There is no loss in generality if we take $a = 1$, and so the standard form of the equation of a circle will be

$$S \equiv x^2 + y^2 + 2gx + 2fy + c = 0. \qquad (18.3)$$

The radius of this circle is $\sqrt{g^2 + f^2 - c}$ and its centre is at $(-g, -f)$.

In a problem, if possible, select the origin as centre and the simpler equation

$$x^2 + y^2 - r^2 = 0$$

now represents the circle with radius r.

In Fig. 13 let PQ be a tangent to the circle at P. Then

$$\begin{aligned} QP^2 = QC^2 - CP^2 &= (x_Q + g)^2 + (y_Q + f)^2 - (g^2 + f^2 - c) \\ &= {x_Q}^2 + {y_Q}^2 + 2gx_Q + 2fy_Q + c. \end{aligned}$$

The tangent to a circle is real if it is drawn from an external point. Hence we see that $S \equiv x^2 + y^2 + 2gx + 2fy + c > 0$ for all points outside the circle, but $S < 0$ for all points inside the circle. In the application of this criterion, it is necessary to ensure that the coefficient of x^2 is positive (unity with our convention that $a = 1$).

Example 1. Find the equation of the circle with centre at $(1, -2)$ and whose radius is 4.

Example 2. Obtain the coordinates of the centre and the radius of the circle represented by $4x^2 + 4y^2 + 4x - 10y - 5 = 0$.

Example 3. Obtain the equation of the circle through the three points $(3, 1)$, $(4, -3)$ and $(1, -1)$.

Example 4. Show that the equation of the circle on the line joining A and B as diameter is $(x - x_A)(x - x_B) + (y - y_A)(y - y_B) = 0$.

19. Conjugate points

Consider the two points A and B, not lying on the circle

$$S \equiv x^2 + y^2 + 2gx + 2fy + c = 0.$$

Then the coordinates of the point P which divides AB in the ratio $\lambda : \mu$ are

$$\left(\frac{\mu x_A + \lambda x_B}{\mu + \lambda}, \frac{\mu y_A + \lambda y_B}{\mu + \lambda}\right).$$

If this point P lies on the circle, we have

$$\left(\frac{\mu x_A + \lambda x_B}{\mu + \lambda}\right)^2 + \left(\frac{\mu y_A + \lambda y_B}{\mu + \lambda}\right)^2 + 2g\left(\frac{\mu x_A + \lambda x_B}{\mu + \lambda}\right) + 2f\left(\frac{\mu y_A + \lambda y_B}{\mu + \lambda}\right) + c = 0.$$

On multiplication by $(\mu + \lambda)^2$, this equation simplifies to

$$\mu^2 S_A + 2\mu\lambda T_{AB} + \lambda^2 S_B = 0, \tag{19.1}$$

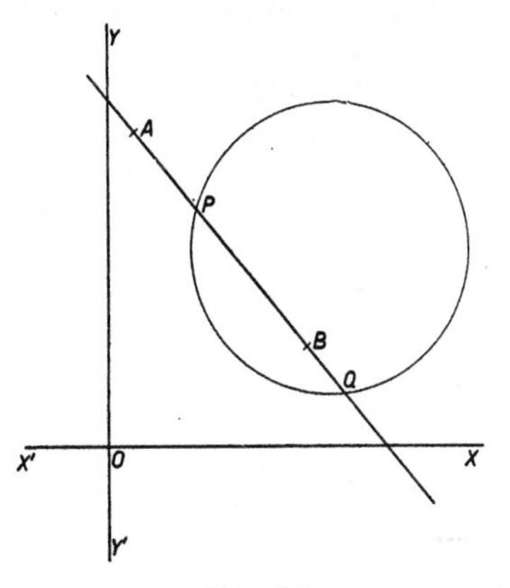

FIG. 14

where

$$S_A \equiv x_A{}^2 + y_A{}^2 + 2gx_A + 2fy_A + c,$$
$$S_B \equiv x_B{}^2 + y_B{}^2 + 2gx_B + 2fy_B + c,$$
$$T_{AB} = T_{BA} \equiv x_A x_B + y_A y_B + g(x_A + x_B) + f(y_A + y_B) + c,$$

Equation (19.1) is called Joachimsthal's quadratic equation and its roots correspond to the two points of intersection P and Q of AB and the circle.

Two points A and B are said to be **conjugate** with respect to the circle if the points P and Q divide A and B harmonically. From §5, we see that in this case the roots of the quadratic in λ/μ are equal in magnitude but opposite in sign. That is, the sum of the roots of (19.1) is zero. Thus the necessary and sufficient condition that the points A and B be conjugate with respect to the circle (18.3) is

$$T_{AB} \equiv x_A x_B + y_A y_B + g(x_A + x_B) + f(y_A + y_B) + c = 0.$$

Example 5. For what value of λ are the points $(2, \lambda)$ and $(\lambda, -1)$ conjugate to the circle $x^2 + y^2 - 2x + 4y - 7 = 0$.

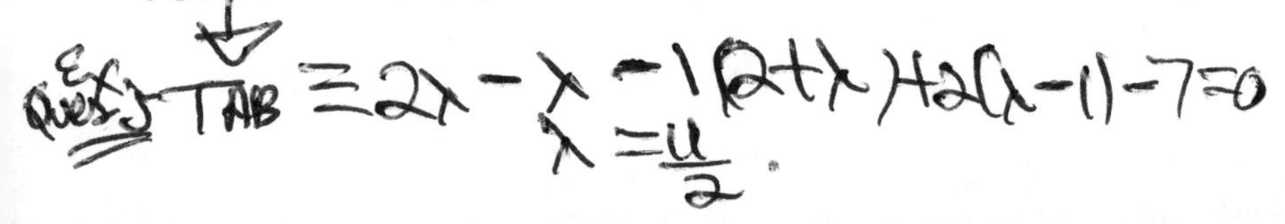

20. Pole and polar

The **polar** of a point A with respect to a circle is defined to be the locus of all points conjugate to A. The point B is conjugate to A with respect to the circle (18.3) if $T_{AB} = 0$. The locus of B is obtained by allowing the coordinates x_B, y_B to become variable; this is achieved by omitting the suffix B. We then find that the locus is given by

$$T_A \equiv x_A x + y_A y + g(x + x_A) + f(y + y_A) + c = 0. \qquad (20.1)$$

This equation is linear in x and y, which shows that the polar of a point is a straight line.

Conversely, the pole of a straight line is defined to be that point whose polar is the given straight line. Let us now obtain the coordinates of the pole A of the straight line $lx + my + n = 0$. The polar of A is given by (20.1), which must then represent the same straight line as $lx + my + n = 0$. The comparison of coefficients yields

$$x_A + g = \rho l; \quad y_A + f = \rho m; \quad gx_A + fy_A + c = \rho n,$$

where ρ is some factor of proportionality. Thus

$$x_A = \rho l - g; \quad y_A = \rho m - f; \quad g(\rho l - g) + f(\rho m - f) + c = \rho n.$$

Therefore

$$\rho = \frac{g^2 + f^2 - c}{gl + fm - n},$$

and we have

$$x = \frac{l(f^2 - c) - mfg + ng}{lg + mf - n}; \quad y = \frac{-lfg + m(g^2 - c) + nf}{lg + mf - n}.$$

This formula is too cumbersome to be quoted. Instead the reader is asked to use the method of this section in relevant examples.

We note that if the circle is $x^2 + y^2 = r^2$, then the two points A and B are conjugate if $x_A x_B + y_A y_B = r^2$, the polar of A is $x_A x + y_A y = r^2$ and the pole of $lx + my + n = 0$ is at the point $(-r^2 l/n, -r^2 m/n)$.

Example 6. Write down the equations of the polars of the points $(-\frac{1}{2}, 0)$, $(-2, -1)$ and $(4, 3)$ with respect to the circle $x^2 + y^2 - 3x + 12y + 6 = 0$ and show that they are concurrent.

Example 7. Obtain the coordinates of the poles of the straight lines $3x - 11y - 13 = 0$, $8x + y - 2 = 0$ and $3x + 2y + 1 = 0$ with respect to the circle $x^2 + y^2 - 4x + 3 = 0$, and show that they are collinear

21. Conjugate lines

Consider the two straight lines $u_1 \equiv l_1x + m_1y + n_1 = 0$ and $u_2 \equiv l_2x + m_2y + n_2 = 0$. The pole of $u_1 = 0$ with respect to the circle (18.3) is, by the preceding section, at the point

$$\left(\frac{(f^2 - c)l_1 - fgm_1 + gn_1}{gl_1 + fm_1 - n_1}, \frac{-fgl_1 + (g^2 - c)m_1 + fn_1}{gl_1 + fm_1 - n_1}\right).$$

The necessary and sufficient condition that this point lies on $u_2 = 0$ is

$$l_2\{(f^2 - c)l_1 - fgm_1 + gn_1\} + m_2\{-fgl_1 + (g^2 - c)m_1 + fn_1\} + n_2\{gl_1 + fm_1 - n_1\} = 0.$$

That is,

$$(c - f^2)l_1l_2 + (c - g^2)m_1m_2 + n_1n_2 + fg(l_1m_2 + l_2m_1) - f(m_1n_2 + m_2n_1) - g(n_1l_2 + n_2l_1) = 0. \quad (21.1)$$

This relation is still valid if the suffices 1 and 2 are interchanged and so it follows that the pole of $u_2 = 0$ also lies on $u_1 = 0$.

When two lines are so situated that each passes through the pole of the other, they are called **conjugate lines.** The necessary and sufficient condition for the conjugacy of the two straight lines $u_1 = 0$ and $u_2 = 0$ is that equation (21.1) be satisfied.

This condition reduces to

$$r^2(l_1l_2 + m_1m_2) = n_1n_2$$

when the centre of the circle is at the origin and r is its radius.

Example 8. Without using (21.1) show that the straight lines $x - y = 0$ and $x - 9y + 6 = 0$ are conjugate with respect to the circle $x^2 + y^2 - 3x + 7y - 5 = 0$. These two straight lines must pass through the poles of each other.

22. Tangent and normal

Let us return to Joachimsthal's equation (19.1), namely

$$\mu^2S_A + 2\mu\lambda T_{AB} + \lambda^2S_B = 0,$$

whose roots correspond to the points of intersection of AB and the circle

$$S \equiv x^2 + y^2 + 2gx + 2fy + c = 0.$$

Suppose A now lies on the circle (Fig. 15); then $S_A = 0$. We define the tangent at A to be the limiting position of the secant AL through A as L tends to coincide with A. Consequently the tangent AB cuts the circle in coincident points at A, and so Joachimsthal's equation $2\mu\lambda T_{AB} + \lambda^2S_B' = 0$ must have coincident roots $\lambda/\mu = 0$.

Therefore, $T_{AB} = 0$ if B lies on the tangent at A. That is, the coordinates of all points on the tangent at A satisfy

$$T_A \equiv x_A x + y_A y + g(x + x_A) + f(y + y_A) + c = 0. \quad (22.1)$$

Hence, this equation represents the tangent at A. It is identical with (20.1) which gives the polar of A. We see then that the pole of a tangent line is at its point of contact.

The **normal** is the straight line through a point on the circumference of a circle, perpendicular to the tangent at this point. Thus the equation of the normal at A is

$$(y_A + f)(x - x_A) - (x_A + g)(y - y_A) = 0.$$

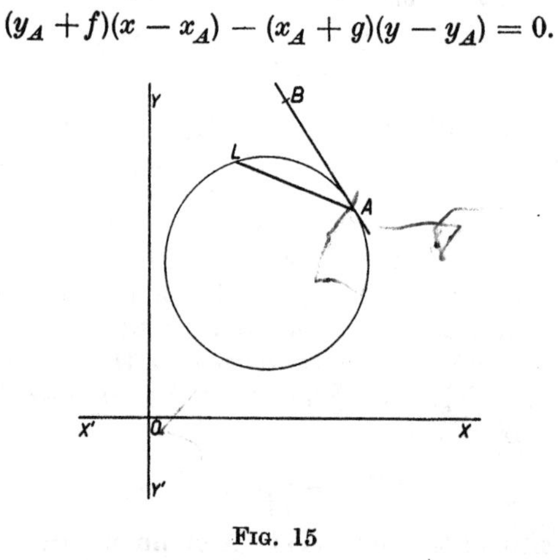

FIG. 15

We verify that the point $(-g, -f)$ lies on this straight line. This is the well-known result that all normals to a circle pass through the centre.

Example 9. Show that the circles $x^2 + y^2 - 2x + 4y + 3 = 0$ and $x^2 + y^2 - 8x - 2y + 9 = 0$ touch one another at $(2, -1)$.

23. Condition that a straight line touch a circle

The pole of a tangent line is its point of contact. Thus a tangent line is conjugate to itself and we say that it is **self-conjugate.** Therefore, we set $l_1 = l_2 = l$ in (21.1) and we obtain that the straight line $lx + my + n = 0$ is a tangent to the circle

$$x^2 + y^2 + 2gx + 2fy + c = 0$$

if

$$(c - f^2)l^2 + (c - g^2)m^2 + n^2 + 2fglm - 2fmn - 2gnl = 0. \quad (23{\cdot}1)$$

Alternatively, the perpendicular distance from the centre of a circle to a tangent line is equal to its radius. This yields

$$\frac{(-lg - mf + n)^2}{l^2 + m^2} = g^2 + f^2 - c,$$

which reduces to (23.1).

If the circle is $x^2 + y^2 = r^2$, the tangency condition simplifies to

$$r^2(l^2 + m^2) = n^2.$$

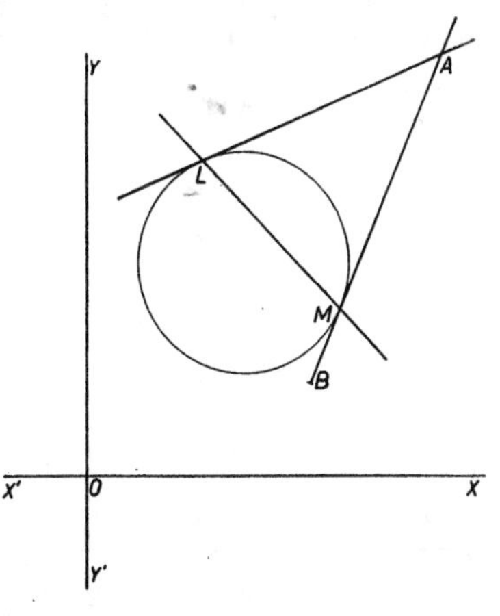

FIG. 16

Example 10. Show that the straight line $y = gx + a\sqrt{1 + g^2}$ touches the circle $x^2 + y^2 = a^2$ for all values of g.

Example 11. For what values of λ does the straight line $4x + \lambda y + 7 = 0$ touch the circle $x^2 + y^2 - 6x + 4y - 12 = 0$.

24. Pair of tangents from a point

Return once more to Joachimsthal's equation (19.1). If AB is a tangent, but neither A nor B lies on the circle, then the roots of (19.1) must coincide. That is $S_A S_B - T_{AB}^2 = 0$. Thus the locus of the point B if A is held fixed is given by

$$S_A S - T_A^2 = 0. \tag{24.1}$$

This is a quadratic equation and must then represent the pair of tangents from A to the circle.

The polar of A, given by $T_A = 0$, cuts the circle $S = 0$ where $T_A = S = 0$. These two points also lie on the line-pair (24.1). Thus the polar of A is the line LM (Fig. 16) joining the points of contact of the tangents from A to the circle.

If A lies inside the circle, the pair of tangents is complex but the polar line LM is real. (Of course, in this case L and M are complex.)

Example 12. Show that the pair of tangents from the point (1, 2) to the circle $x^2 + y^2 - 4x + 2y = 0$ are mutually orthogonal.

25. Angle of intersection of two circles

The angle of intersection of two circles is the angle between the two tangents at a point of intersection. It is a geometrical fact that this angle is the same at both intersections A and B, Fig. 17. If

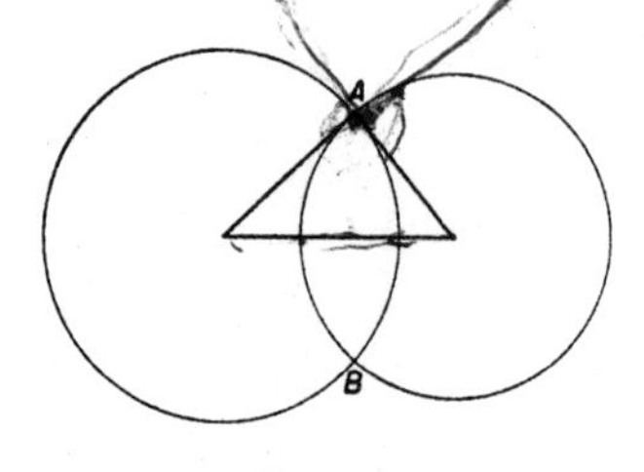

FIG. 17

this angle is a right angle the two circles are said to cut orthogonally. Then each tangent passes through the centre of the other and it follows that the square of the distance between their two centres is equal to the sum of the squares of the two radii. Let the circles be

$$x^2 + y^2 + 2g_1x + 2f_1y + c_1 = 0,$$
$$x^2 + y^2 + 2g_2x + 2f_2y + c_2 = 0.$$

Then these circles cut orthogonally if

$$(g_1 - g_2)^2 + (f_1 - f_2)^2 = g_1^2 + f_1^2 - c_1 + g_2^2 + f_2^2 - c_2.$$

That is,

$$2g_1g_2 + 2f_1f_2 - c_1 - c_2 = 0. \tag{25.1}$$

Example 13. Calculate the angle of intersection of the two circles $8x^2 + 8y^2 + 16x + 3 = 0$ and $8x^2 + 8y^2 + 16y + 3 = 0$.

26. Radical axis

Consider the two circles

$$S_1 \equiv x^2 + y^2 + 2g_1x + 2f_1y + c_1 = 0,$$
$$S_2 \equiv x^2 + y^2 + 2g_2x + 2f_2y + c_2 = 0.$$

The square of the length of the tangent from P to the circle $S_1 = 0$ is $x_P^2 + y_P^2 + 2g_1x_P + 2f_1y_P + c$ (see §18). Thus the locus of points $P = (x, y)$ from which the tangents to $S_1 = 0$ and $S_2 = 0$ are equal is given by

$$x^2 + y^2 + 2g_1x + 2f_1y + c_1 = x^2 + y^2 + 2g_2x + 2f_2y + c_2.$$

That is,

$$S_1 - S_2 \equiv 2(g_1 - g_2)x + 2(f_1 - f_2)y + c_1 - c_2 = 0. \quad (26.1)$$

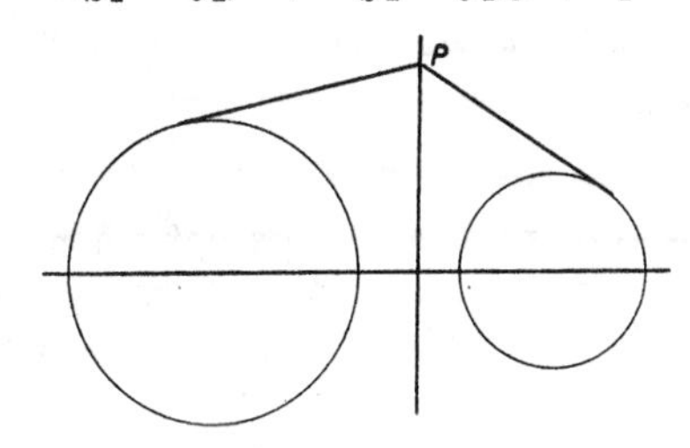

FIG. 18 (a)

This equation is linear and so represents a straight line, called the **radical axis** of the two circles. The gradient of the radical axis is $-(g_1 - g_2)/(f_1 - f_2)$ and so it is orthogonal to the line of gradient $(f_1 - f_2)/(g_1 - g_2)$ joining the centres of the two circles. The radical

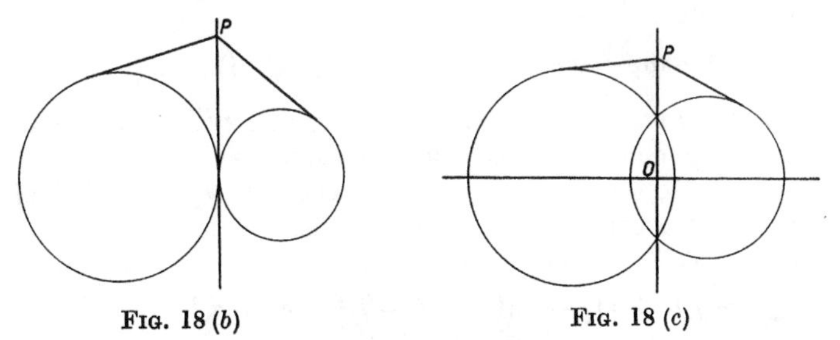

FIG. 18 (b) FIG. 18 (c)

axis obviously passes through any common points the two circles may have. Figs. 18a, 18b and 18c depict the radical axis in the respective cases of non-intersecting, touching and intersecting circles.

Let us introduce a third circle

$$S_3 \equiv x^2 + y^2 + 2g_3x + 2f_3y + c_3 = 0.$$

We now have the three radical axes $S_1 - S_2 = 0$, $S_2 - S_3 = 0$ and $S_3 - S_1 = 0$. These lines are concurrent at the point $S_1 = S_2 = S_3$. This point of concurrence is called the **radical centre** of the three circles.

Example 14. Find the coordinates of the radical centre of the three circles $x^2 + y^2 - 4x - 2y + 6 = 0$, $x^2 + y^2 - 2x + 6y = 0$ and $x^2 + y^2 - 12x$

$+ 2y + 30 = 0$. Further, show that these circles all cut the circle $x^2 + y^2 - 6x + 6 = 0$ orthogonally.

27. Coaxal circles

Consider two circles

$$S_1 \equiv x^2 + y^2 + 2g_1x + 2f_1y + c_1 = 0,$$
$$S_2 \equiv x^2 + y^2 + 2g_2x + 2f_2y + c_2 = 0,$$

and set up the equation

$$S_1 + \lambda S_2 = 0. \tag{27.1}$$

This represents a system of circles, except when $\lambda = -1$, which corresponds to the radical axis of $S_1 = 0$ and $S_2 = 0$. The radical axis of the two *distinct* circles for which $\lambda = \lambda_1$ and $\lambda = \lambda_2$ ($\lambda_1 \neq \lambda_2$) is

$$\frac{S_1 + \lambda_1 S_2}{1 + \lambda_1} - \frac{S_1 + \lambda_2 S_2}{1 + \lambda_2} = 0.$$

On multiplication by $(1 + \lambda_1)(1 + \lambda_2)$ and subsequent division by $(\lambda_1 - \lambda_2)$ which is not zero, this equation reduces to $S_1 - S_2 = 0$. Thus (27.1) represents a system of circles, called a **coaxal system** of circles, with a common radical axis. The centre of the circle (27.1) is at the point

$$(-(g_1 + \lambda g_2)/(1 + \lambda), -(f_1 + \lambda f_2)/(1 + \lambda)). \tag{27.2}$$

Therefore, the centres of all circles of a coaxal system are collinear on a line perpendicular to the common radical axis.

Let us choose the common radical axis as the y-axis and the line of centres orthogonal to it as the x-axis. From (26.1) we have $f_1 = f_2$ and $c_1 = c_2$. The point with coordinates (27.2) now lies on the x-axis, and so $f_1 = f_2 = 0$. Accordingly we may write (27.1), on division by $(1 + \lambda)$ in the standard form

$$x^2 + y^2 + 2\mu x + c = 0, \tag{27.3}$$

where we have put $\mu = (g_1 + \lambda g_2)/(1 + \lambda)$ and $c_1 = c_2 = c$. This equation may be written

$$(x + \mu)^2 + y^2 = \mu^2 - c$$

and there are three cases to be considered:

(1) $c < 0$. On setting $c = -k^2$, we see that all circles of the coaxal system pass through the two fixed points $P \equiv (0, k)$ and $Q \equiv (0, -k)$ of Fig. 19a. The circle on PQ as diameter is the smallest circle of the system.

(2) $c = 0$. In this case, all circles of the system touch the y-axis at the origin as in Fig. 19b.

(3) $c > 0$. Let us write $c = k^2$ and it follows that the circles of the coaxal system are real if and only if $-k < \mu < k$. In particular, for $\mu = \pm k$ there are two point circles of the system $R \equiv (k, 0)$ and $S \equiv (-k, 0)$ (Fig. 19c), called **limiting points.** The radical axis $x = 0$ cuts the coaxal system where $y^2 + k^2 = 0$ and so the points of intersection are not real in this case.

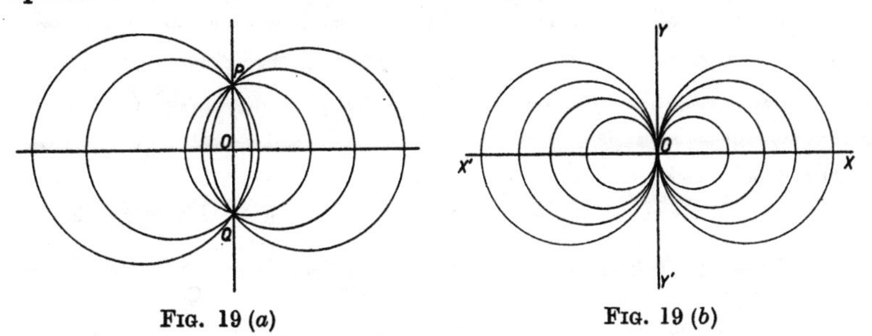

FIG. 19 (a) FIG. 19 (b)

Example 15. Obtain the limiting points of the coaxal system of circles determined by $x^2 + y^2 + 2x + 5 = 0$ and $x^2 + y^2 + 2y + 5 = 0$.

Example 16. Find the equation of the circle through the points of intersection of the circles $x^2 + y^2 = 1$ and $x^2 + y^2 + 2x + 4y + 1 = 0$, which touches the straight line $x + 2y + 5 = 0$.

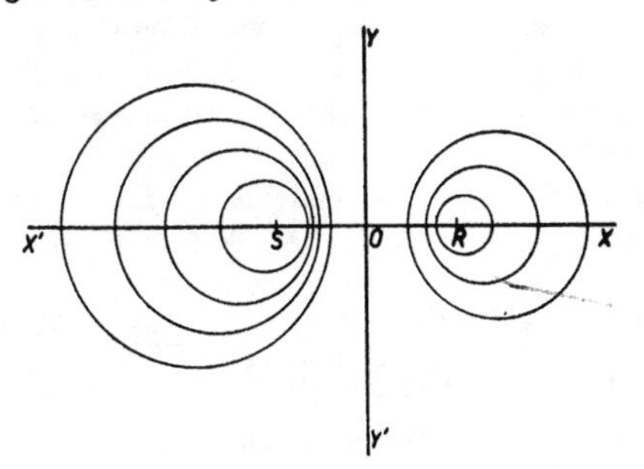

FIG. 19 (c)

28. Conjugate system of coaxal circles

If the circle

$$S \equiv x^2 + y^2 + 2gx + 2fy + d = 0$$

cuts the two circles of a coaxal system

$$S_1 \equiv x^2 + y^2 + 2\mu_1 x + c = 0$$
$$S_2 \equiv x^2 + y^2 + 2\mu_2 x + c = 0$$

orthogonally, then by (25.1) we have

$$g\mu_1 - c - d = 0; \quad g\mu_2 - c - d = 0.$$

Hence $g = 0$ and $d = -c$, and so all circles of the coaxal system

$$x^2 + y^2 + 2\mu x + c = 0$$

cuts all circles of the coaxal system

$$x^2 + y^2 + 2\nu y - c = 0$$

orthogonally. These two coaxal systems are said to be ***conjugate.***

Miscellaneous examples

1. Find the locus of a point which moves so that its polars with respect to two fixed circles are mutually orthogonal.

2. If the four points of intersection of the straight lines $l_1x + m_1y + n_1 = 0$ and $l_2x + m_2y + n_2 = 0$ and the circles $x^2 + y^2 + 2g_1x + 2f_1y + c_1 = 0$, $x^2 + y^2 + 2g_2x + 2f_2y + c_2 = 0$ respectively are concyclic, show that $2(g_1 - g_2)(m_1n_2 - m_2n_1) + 2(f_1 - f_2)(n_1l_2 - n_2l_1) + (c_1 - c_2)(l_1m_2 - l_2m_1) = 0$.

3. Obtain the condition that the circle $x^2 + y^2 + 2g_1x + 2f_1y + c_1 = 0$ should cut the circle $x^2 + y^2 + 2g_2x + 2f_2y + c_2 = 0$ at the ends of a diameter. Hence deduce that in general two circles of a coaxal system are cut at the ends of a diameter by a given circle but only one circle of a coaxal system cuts a given circle at the ends of a diameter.

4. Without solving for the points of intersection of the two circles $x^2 + y^2 + 18x - 2y - 15 = 0$ and $x^2 + y^2 - 6x - 4y - 5 = 0$ obtain the equation of the circle whose diameter is their common chord.

5. If P and Q are two points outside a circle, prove that the sum of the squares of the tangents from P and Q to the circle is equal to PQ^2 if P and Q are conjugate points.

6. Show that the locus of the mid-points of the chords of the circle $x^2 + y^2 + 2gx + c = 0$ which pass through the origin is the circle $x^2 + y^2 + gx = 0$.

7. Show that $x^2 + y^2 + 2gx + 2fy + c + \lambda(lx + my + n) = 0$ represents a coaxal system of circles as λ varies, with radical axis $lx + my + n = 0$.

8. Obtain the equations of the circumcircle and the nine-points circle of the triangle formed by the points $R \equiv (-\lambda, 0)$, $S \equiv (\lambda, 0)$ and $T \equiv (\alpha, \beta)$. Further find the equation of the radical axis of these two circles.

9. Obtain the locus of the poles of the fixed straight line $\frac{x}{a} + \frac{y}{b} = 1$ with respect to circles in the first quadrant which touch both coordinate axes. When does this locus degenerate to a straight line? Interpret geometrically.

10. Show that the limiting points of the coaxal system of circles $x^2 + y^2 + 2gx + c + \lambda(x^2 + y^2 + 2fy + c') = 0$ are real if $(c - c')^2 > 4(f^2g^2 - f^2c - g^2c')$.

Answers

1. $x^2 + y^2 - 2x + 4y - 11 = 0$. **2.** $(-\frac{1}{2}, \frac{5}{4})$, $7/4$. **3.** $5x^2 + 5y^2 - 31x + 11y + 32 = 0$. **5.** $11/2$. **6.** $8x - 24y - 27 = 0$, $7x - 10y - 6 = 0$, $5x + 18y +$

36 $= 0$. **7.** $(17/7, -11/7)$, $(10/7, -1/14)$, $(11/7, -2/7)$. **11.** -3 or $-13/21$. **13.** $\tan^{-1}(4/3)$. **14.** $(3, 0)$. **15.** $(1, -2)$, $(-2, 1)$. **16.** $x^2 + y^2 + x + 2y = 0$.

Miscellaneous examples. **1.** The circle on the join of centres as diameter. **3.** $2(g_2^2 + f_2^2 - g_1g_2 - f_1f_2) + c_1 - c_2 = 0$. **4.** $145(x^2 + y^2) - 78x - 514y - 1055 = 0$. **8.** $x^2 + y^2 - (\alpha^2 + \beta^2 - \lambda^2)y/\beta - \lambda^2 = 0$; $x^2 + y^2 - \alpha x + (\alpha^2 - \beta^2 - \lambda^2)y/2\beta = 0$; $\alpha x - (3\alpha^2 + \beta^2 - 3\lambda^2)y/2\beta - \lambda^2 = 0$. **9.** $(ax - by)(bx - ay) - ab(a - b)(x - y) = 0$; $a = b$.

Miscellaneous examples

Ques. 4

Ques. 4, P32

(a) $y = -12x + 5$

(b) $x = -\frac{1}{12}y + \frac{5}{12}$

Substitute these values for x and y into

(1) $x^2 + y^2 + 18x - 2y - 15 = 0$

to get $145x^2 + 145y^2 - 78x - 514y - 1055 = 0$.

(a) + (b) values, Subtract

(2) $x^2 + y^2 - 6x - 4y - 5 = 0$ from

(1) $x^2 + y^2 + 18x - 2y - 15 = 0$

to get $y + 12x - 5 = 0$.

CHAPTER IV

ELLIPSE

29. Ellipse

An **ellipse** is defined to be the locus of the point P (Fig. 20) such that $PM^2 = k^2 A'M \,.\, MA$, where A and A' are two fixed points and PM is perpendicular to $A'A$. Let us choose the mid-point O

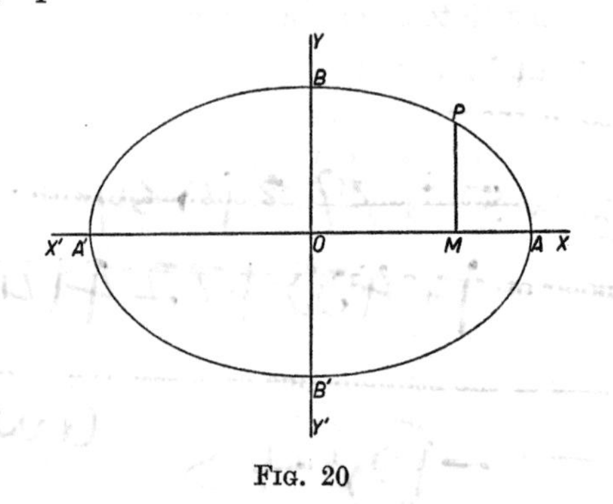

FIG. 20

of $A'A$ as origin and OA as the direction of the positive x-axis. Then from the definition it follows that

$$y^2 = k^2(x + a)(a - x) = k^2(a^2 - x^2)$$

if $A'A = 2a$. Introduce $b^2 = k^2a^2$ and the equation of the ellipse takes the standard form

$$\frac{x^2}{a^2} + \frac{y^2}{b^2} - 1 = 0. \tag{29.1}$$

If $a > b$, $A'A$ is called the **major axis.** The ellipse cuts the y-axis at the points $B \equiv (0, b)$ and $B' \equiv (0, -b)$. Their join BB' is called the **minor axis.** Further, the points A, A', B and B' are called **vertices,** whilst O is called the **centre.** If (x, y) lies on the ellipse, so does $(-x, y)$, $(x, -y)$ and $(-x, -y)$. Hence the ellipse is clearly symmetrical about both the major and minor axes. Also any chord through the centre, called a **diameter,** is bisected there.

If $a = b$, the ellipse becomes a circle. This would correspond to $k = 1$.

If $a < b$, we call $A'A$ and $B'B$ the minor and major axis respectively.

30. Tangent and polar properties

The theory centred around Joachimsthal's quadratic equation (19.1) can be repeated for the ellipse (29.1) with the modifications

$$S_A \equiv \frac{x_A{}^2}{a^2} + \frac{y_A{}^2}{b^2} - 1; \quad T_A = \frac{x_A x}{a^2} + \frac{y_A y}{b^2} - 1;$$

$$T_{AB} = \frac{x_A x_B}{a^2} + \frac{y_A y_B}{b^2} - 1,$$

and the reader is advised to work out the following results corresponding to sections 19 to 24 on the circle:

(i) two points A and B are conjugate with respect to the ellipse, if the intersections P and Q of AB and the ellipse divide A and B harmonically. The necessary and sufficient condition that A and B be conjugate with respect to the ellipse is $T_{AB} = 0$;

(ii) the polar of A, defined to be the locus of all points conjugate to A is the straight line $T_A = 0$;

(iii) the pole of the straight line $lx + my + n = 0$ with respect to the ellipse, defined to be that point whose polar is $lx + my + n = 0$, is at the point $(-a^2l/n, -b^2m/n)$;

(iv) the two straight lines $l_1x + m_1y + n_1 = 0$, $l_2x + m_2y + n_2 = 0$ are conjugate, that is each passes through the pole of the other if

$$a^2l_1l_2 + b^2m_1m_2 = n_1n_2;$$

(v) the tangent at A, defined as the limiting position of a secant as its points of intersection with the ellipse tend to coincidence, is given by $T_A = 0$;

(vi) the normal, defined to be the straight line perpendicular to the tangent at the point of contact, at A has the equation

$$\frac{y_A}{b^2}(x - x_A) - \frac{x_A}{a^2}(y - y_A) = 0;$$

(vii) the straight line $lx + my + n = 0$ touches the ellipse if

$$a^2l^2 + b^2m^2 = n^2;$$

(viii) the pair of tangents from A to the ellipse is given by $S_AS - T_A{}^2 = 0$;

(ix) the polar of A is the line joining the points of contact of the tangents from A to the ellipse.

Example 1. By examining the reality of the pair of tangents from the point A to the ellipse $\frac{x^2}{a^2} + \frac{y^2}{b^2} - 1 = 0$ show that A lies inside or outside the ellipse according as $S_A < 0$ or $S_A > 0$.

Example 2. Show that the straight line $y = gx + \sqrt{a^2g^2 + b^2}$ touches the ellipse $\frac{x^2}{a^2} + \frac{y^2}{b^2} - 1 = 0$ for all values of g.

Example 3. Find the coordinates of the mid-point of the chord $lx + my + n = 0$ of the ellipse $\frac{x^2}{a^2} + \frac{y^2}{b^2} - 1 = 0$.

31. Orthoptic locus

Let us find the locus of the points of intersection of perpendicular tangents to the ellipse $S \equiv \frac{x^2}{a^2} + \frac{y^2}{b^2} - 1 = 0$. The pair of tangents from A to the ellipse is (see §30, viii) given by $S_AS - T_A{}^2 = 0$. That is,

$$\left(\frac{x_A{}^2}{a^2} + \frac{y_A{}^2}{b^2} - 1\right)\left(\frac{x^2}{a^2} + \frac{y^2}{b^2} - 1\right) - \left(\frac{x_Ax}{a^2} + \frac{y_Ay}{b^2} - 1\right)^2 = 0.$$

This line pair is mutually orthogonal if the sum of the coefficients of x^2 and y^2 is zero (see §16). Thus

$$\left(\frac{1}{a^2} + \frac{1}{b^2}\right)\left(\frac{x_A{}^2}{a^2} + \frac{y_A{}^2}{b^2} - 1\right) - \frac{x_A{}^2}{a^4} - \frac{y_A{}^2}{b^4} = 0,$$

which reduces to

$$x_A{}^2 + y_A{}^2 = a^2 + b^2.$$

Accordingly the locus of A is the circle

$$x^2 + y^2 = a^2 + b^2,$$

called the **orthoptic** circle. Its centre coincides with the centre of the ellipse.

Example 4. Obtain the equation of the orthoptic circle by considering the intersection of the perpendicular tangents at two points on the ellipse.

32. Parametric Equations

Draw the circle, called the auxiliary circle on the major axis $A'A$ as diameter (Fig. 21). Let the ordinate through P cut this circle at Q. Introduce the angle $AOQ = \theta$. Then

$$x_P = OM = OQ \cos\theta = a\cos\theta.$$

Substitute in the equation (29.1) of the ellipse and the result is that $y_P = b\sin\theta$.

If we allow θ, called the **eccentric angle,** to vary from 0 to 2π the point P travels once round the ellipse in the counter-clockwise direction starting at A and finishing at A. Consequently we may say that the ellipse has the parametric equations

$$x = a \cos \theta, \quad y = b \sin \theta.$$

Occasionally we shall refer to the "point θ" on the ellipse, meaning the point P whose eccentric angle is θ.

The equation of the chord joining the two points α and β is

$$\frac{y - b \sin \alpha}{b \sin \alpha - b \sin \beta} = \frac{x - a \cos \alpha}{a \cos \alpha - a \cos \beta}.$$

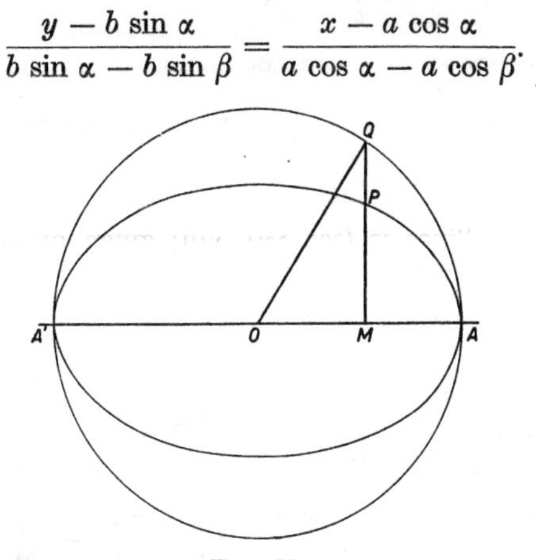

FIG. 21

That is†

$$\frac{y - b \sin \alpha}{2b \cos \frac{1}{2}(\alpha + \beta) \sin \frac{1}{2}(\alpha - \beta)} = \frac{x - a \cos \alpha}{2a \sin \frac{1}{2}(\alpha + \beta) \sin \frac{1}{2}(\beta - \alpha)}$$

which reduces to

$$\frac{x}{a} \cos \tfrac{1}{2}(\alpha + \beta) + \frac{y}{b} \sin \tfrac{1}{2}(\alpha + \beta) = \cos \tfrac{1}{2}(\alpha - \beta).$$

---

† The reader is advised to familiarize himself with the following trigono metrical formulae:

$$\cos \alpha + \cos \beta = 2 \cos \tfrac{1}{2}(\alpha + \beta) \cos \tfrac{1}{2}(\alpha - \beta);$$
$$\cos \alpha - \cos \beta = 2 \sin \tfrac{1}{2}(\alpha + \beta) \sin \tfrac{1}{2}(\beta - \alpha);$$
$$\sin \alpha + \sin \beta = 2 \sin \tfrac{1}{2}(\alpha + \beta) \cos \tfrac{1}{2}(\alpha - \beta);$$
$$\sin \alpha - \sin \beta = 2 \cos \tfrac{1}{2}(\alpha + \beta) \sin \tfrac{1}{2}(\alpha - \beta);$$
$$2 \sin \alpha \cos \beta = \sin (\alpha + \beta) + \sin (\alpha - \beta);$$
$$2 \cos \alpha \sin \beta = \sin (\alpha + \beta) - \sin (\alpha - \beta);$$
$$2 \sin \alpha \sin \beta = \cos (\alpha - \beta) - \cos (\alpha + \beta);$$
$$2 \cos \alpha \cos \beta = \cos (\alpha + \beta) + \cos (\alpha - \beta).$$

The equation of the tangent at the point θ follows from §30 (v) or from the above equation by making $\alpha = \beta = \theta$ and the result is

$$\frac{x}{a} \cos \theta + \frac{y}{b} \sin \theta = 1.$$

Example 5. Show that the tangents at the points α and β on the ellipse $x = a \cos \theta$, $y = b \sin \theta$ intersect at the point $(a \cos \frac{1}{2}(\alpha + \beta)/\cos \frac{1}{2}(\alpha - \beta)$, $b \sin \frac{1}{2}(\alpha + \beta)/\cos \frac{1}{2}(\alpha - \beta))$.

Example 6. Prove that the normals at the points α and β on the ellipse $x = a \cos \theta$, $y = b \sin \theta$ intersect at the point $((a^2 - b^2) \cos \alpha \cos \beta \cos \frac{1}{2}(\alpha + \beta)/a \cos \frac{1}{2}(\alpha - \beta)$, $-(a^2 - b^2) \sin \alpha \sin \beta \sin \frac{1}{2}(\alpha + \beta)/b \cos \frac{1}{2}(\alpha - \beta))$.

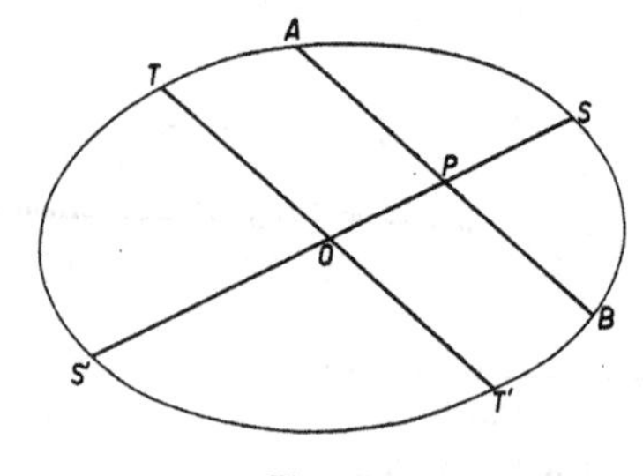

FIG. 22

Example 7. Prove that $lx + my + n = 0$ is a normal to the ellipse $\frac{x^2}{a^2} + \frac{y^2}{b^2} - 1 = 0$ if $\frac{a^2}{l^2} + \frac{b^2}{m^2} = \frac{(a^2 - b^2)^2}{n^2}$.

33. Conjugate diameters

We now seek the locus of the mid-points of a system of parallel chords of the ellipse $x = a \cos \theta$, $y = b \sin \theta$. Let A and B (Fig. 22) be the points α and β respectively on the ellipse. Their mid-point $P \equiv (x, y)$ then has the co-ordinates

$$x = \frac{a}{2} (\cos \alpha + \cos \beta) = a \cos \tfrac{1}{2}(\alpha + \beta) \cos \tfrac{1}{2}(\alpha - \beta),$$

$$y = \frac{b}{2} (\sin \alpha + \sin \beta) = f \sin \tfrac{1}{2}(\alpha + \beta) \cos \tfrac{1}{2}(\alpha - \beta),$$

from which we have

$$\frac{y}{x} = \frac{b}{a} \tan \tfrac{1}{2}(\alpha + \beta). \tag{33.1}$$

The equation of the chord AB is

$$\frac{x}{a}\cos\tfrac{1}{2}(\alpha+\beta)+\frac{y}{b}\sin\tfrac{1}{2}(\alpha+\beta)=\cos\tfrac{1}{2}(\alpha-\beta),$$

and its gradient is

$$g=-\frac{b}{a}\cot\tfrac{1}{2}(\alpha+\beta). \tag{33.2}$$

For a system of parallel chords, the gradient g is a constant and so the elimination of $(\alpha+\beta)$ between (33.1) and (33.2) yields

$$\frac{y}{x}=-\frac{b^2}{a^2g}.$$

Let us introduce its gradient $g'=-b^2/a^2g$ and it follows that the locus of P is the diameter $y=g'x$. In Fig. 22, $S'OS$ is this diameter.

If we interchange g and g', it is easy to see that the locus of the mid-points of all chords parallel to $S'OS$ is the diameter $T'OT$ in the figure, whose gradient is g. The diameters $S'OS$ and $T'OT$, each of which bisects all chords parallel to the other, are called **conjugate** diameters. The name conjugate is justified because they satisfy the condition §30 (iv) for conjugate lines with respect to the ellipse. However, the pole of a diameter is not at a finite point.

The product of the gradients g and g' of conjugate diameters is

$$gg'=-\frac{b^2}{a^2}. \tag{33.3}$$

Suppose S and T are the points θ and φ on the ellipse. Then $\dfrac{b\cos\theta}{a\sin\theta}\cdot\dfrac{b\cos\varphi}{a\sin\varphi}=-\dfrac{b^2}{a^2}$, which reduces to $\cos(\theta-\varphi)=0$. Accordingly $|\theta-\varphi|=\pi/2$. Hence the eccentric angles at the ends of two conjugate diameters differ by $\pi/2$. We may then take

$$S\equiv(a\cos\theta,b\sin\theta)\quad\text{and}\quad T\equiv(-a\sin\theta,b\cos\theta).$$

Example 8. Obtain the equation of the chord of the ellipse $\dfrac{x^2}{a^2}+\dfrac{y^2}{b^2}-1=0$ which has its mid-point at A.

Example 9. If $S'OS$ and $T'OT$ are conjugate diameters of the ellipse $\dfrac{x^2}{a^2}+\dfrac{y^2}{b^2}-1=0$, prove that (i) $OS^2+OT^2=a^2+b^2$, (ii) $OS=OT$ if the eccentric angle of either S or T is $\pi/4$, (iii) the area of the parallelogram formed by the tangents at the ends of conjugate diameters is $4ab$.

34. Focus and directrix

We shall see that there are two special points on the *major* axis of the ellipse $\frac{x^2}{a^2} + \frac{y^2}{b^2} - 1 = 0$ which possess remarkable properties. These points F and F', called the **foci**, are distant a from the vertices B and B' (Fig. 23). The number e, called the **eccentricity**, is introduced by the relation

$$e^2 = 1 - \frac{b^2}{a^2} \quad \text{or} \quad b^2 = a^2(1 - e^2).$$

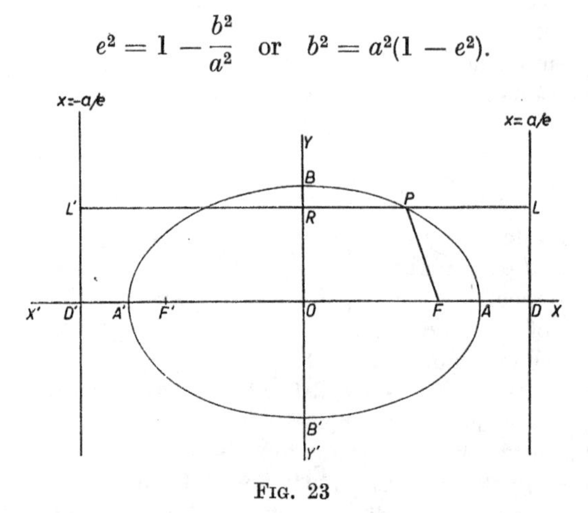

FIG. 23

Since $OF^2 = a^2 - b^2$, it follows that $F \equiv (ae, 0)$ and $F' \equiv (-ae, 0)$. These foci are real points because $b < a$ and this ensures that e is positive and less than unity. (We regard e as an essentially positive number and never select e negative.) The polar of F, called the **directrix**, is the straight line $x = a/e$. From P, the point θ on the ellipse, draw PL perpendicular to the directrix. Then

$$PL = OD - RP = \frac{a}{e} - a \cos \theta = \frac{a}{e}(1 - e \cos \theta),$$

$$\begin{aligned} PF^2 &= (a \cos \theta - ae)^2 + b^2 \sin^2 \theta \\ &= a^2 \cos^2 \theta - 2a^2 e \cos \theta + a^2 e^2 + a^2(1 - e^2) \sin^2 \theta \\ &= a^2 - 2a^2 e \cos \theta + a^2 e^2 \cos^2 \theta \\ &= a^2(1 - e \cos \theta)^2. \end{aligned}$$

Hence $PF/PL = e$. Likewise, introduce a second directrix as the polar of the focus $F' \equiv (-ae, 0)$ and its equation is $x = -a/e$. Then

$$L'P = D'O + RP = \frac{a}{e} + a \cos \theta = \frac{a}{e}(1 + e \cos \theta),$$

$$PF'^2 = (a \cos \theta + ae)^2 + b^2 \sin^2 \theta$$
$$= a^2 \cos^2 \theta + 2a^2e \cos \theta + a^2e^2 + a^2(1 - e^2) \sin^2 \theta$$
$$= a^2(1 + e \cos \theta)^2.$$

Therefore, $PF'/L'P = e$.

Thus an ellipse can be defined as the locus of a point which moves so that its distance from a fixed point, called the focus, is in a constant ratio less than unity to its distance from a fixed line called the directrix. We have just demonstrated that an ellipse has two foci and two directrices.

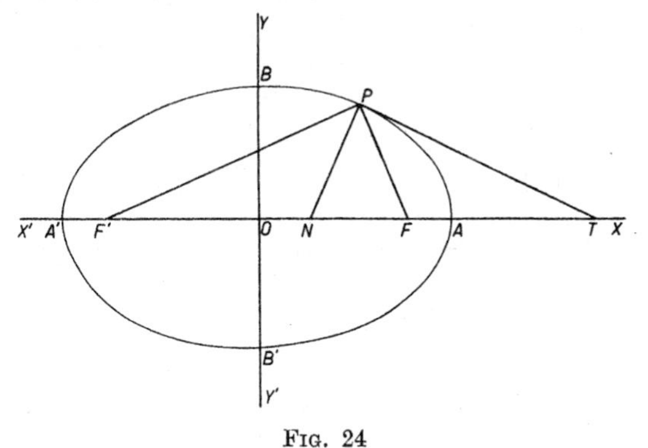

Fig. 24

The chord through a focus perpendicular to the major axis is called the **latus rectum.**

Our calculations have shown us that $PF + PF' = 2a$. This property can be used to define an ellipse as the locus of a point which moves so that the sum of its distances from two fixed points is constant. Hence we can describe an ellipse mechanically by attaching the two ends of a thread of length $2a$ to two fixed points and move a pencil so that the thread is always taut. The pencil point will describe an ellipse with foci at the two fixed points and with major axis $2a$.

Let the tangent at $P \equiv (a \cos \theta, b \sin \theta)$ cut the major axis (Fig. 24) at T. The equation of the tangent at P is $\frac{x}{a} \cos \theta + \frac{y}{b} \sin \theta = 1$ and so $OT = a/\cos \theta$. Thus $FT = \frac{a}{\cos \theta} - ae = \frac{a(1 - e \cos \theta)}{\cos \theta}$ and $F'T = \frac{a}{\cos \theta} + ae = \frac{a(1 + e \cos \theta)}{\cos \theta}$. Therefore,

$$\frac{F'T}{FT} = \frac{1 + e \cos \theta}{1 - e \cos \theta} = \frac{F'P}{FP}.$$

Consequently PT is the external bisector of the angle $F'PF$, and so the normal PN is the internal bisector of the angle between the lines joining a point on the ellipse to the two foci.

Example 10. Show that the length of the semi-latus rectum is $l = b^2/a = a(1 - e^2)$. Hence show that $l/PF = 1 + e \cos \varphi$, where φ is the angle PFT in Fig. 24. (This relation is very important in dynamics.)

Example 11. Let the perpendicular from P to the x-axis, Fig. 24, be PG and prove (i) $ON = e^2 OG$, (ii) $OG \,.\, OT = a^2$.

Example 12. In Fig. 24, prove that the product $PF \,.\, PF'$ is equal to the square on the semi-diameter conjugate to OP.

Example 13. If θ and φ are the eccentric angles of two points collinear with a focus of an ellipse, show that $\cos \frac{1}{2}(\theta - \varphi) = \pm e \cos \frac{1}{2}(\theta + \varphi)$.

Example 14. From a point P on a directrix of an ellipse a pair of tangents PQ and PR are drawn to an ellipse. Prove that Q and R are collinear with the corresponding focus of the ellipse.

Example 15. The lines joining the point P to the foci F and F' of an ellipse cut it again at A and B. Prove that the tangents at A and B and the normal at P are concurrent.

35. Normal

The tangent at the point θ on the ellipse $x = a \cos \theta$, $y = b \sin \theta$ is $\frac{x}{a} \cos \theta + \frac{y}{b} \sin \theta = 1$ and so the normal at θ is

$$\frac{\sin \theta}{b}(x - a \cos \theta) - \frac{\cos \theta}{a}(y - b \sin \theta) = 0,$$

which simplifies to

$$\frac{x}{b} \sin \theta - \frac{y}{a} \cos \theta = \frac{a^2 - b^2}{ab} \sin \theta \cos \theta.$$

Let A be a *fixed* point on the normal at the point θ, then

$$ax_A \sin \theta - by_A \cos \theta - (a^2 - b^2) \sin \theta \cos \theta = 0. \quad (35.1)$$

We may now ask: "How many normals pass through A?" The answer depends on the number of solutions (35.1) has considered as an equation in θ. To solve this equation, we make the substitution $\tan \frac{1}{2}\theta = t$, from which $\sin \theta = 2t/(1 + t^2)$ and $\cos \theta = (1 - t^2)/(1 + t^2)$. A straightforward calculation yields the quartic equation

$$by_A t^4 + 2(ax_A + a^2 - b^2)t^3 + 2(ax_A - a^2 + b^2)t - by_A = 0.$$

Thus, in general, four normals (Fig. 25) can be drawn from a point to an ellipse.

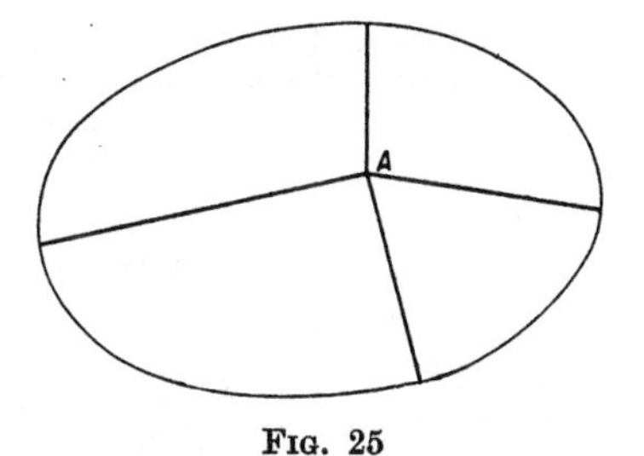

FIG. 25

Let the roots of this quartic be t_1, t_2, t_3 and t_4 corresponding to the eccentric angles θ_1, θ_2, θ_3 and θ_4 respectively. From the theory of equations $t_1t_2t_3t_4 = -1$ and $\Sigma t_1t_2 = 0$. But $\tan \frac{1}{2}(\theta_1 + \theta_2 + \theta_3 + \theta_4) = (\Sigma t_1 - \Sigma t_1t_2t_3)/(1 - \Sigma t_1t_2 + t_1t_2t_3t_4)$ and so $\theta_1 + \theta_2 + \theta_3 + \theta_4 = (2n + 1)\pi$, where n is an integer. This condition is necessary but not sufficient, that the normals at θ_1, θ_2, θ_3, θ_4 be concurrent. Sufficient conditions are $t_1t_2t_3t_4 = -1$ and $\Sigma t_1t_2 = 0$.

36. Concyclic points

To find the condition that the four points θ_1, θ_2, θ_3 and θ_4 of the ellipse $x = a \cos \theta$, $y = b \sin \theta$ be concyclic, suppose that they lie on the circle

$$x^2 + y^2 + 2gx + 2fy + c = 0.$$

The point $(a \cos \theta, b \sin \theta)$ lies on this circle if

$$a^2 \cos^2 \theta + b^2 \sin^2 \theta + 2ga \cos \theta + 2fb \sin \theta + c = 0,$$

and the substitution $\tan \frac{1}{2}\theta = t$ yields the quartic equation

$$(a^2 - 2ga + c)t^4 + 4fbt^3 + 2(-a^2 + 2b^2 + c)t^2 + 4fbt + (a^2 + 2ga + c) = 0.$$

Hence $\Sigma t_1t_2t_3 = \Sigma t_1$ and so $\tan \frac{1}{2}(\theta_1 + \theta_2 + \theta_3 + \theta_4) = 0$ (see previous section). Consequently $\theta_1 + \theta_2 + \theta_3 + \theta_4 = 2n\pi$, where n is an integer. This condition is both necessary and sufficient.

Miscellaneous examples

1. Show that the area of the triangle with vertices at the points α, β and γ of the ellipse $x = a \cos \theta$, $y = b \sin \theta$ is $2ab \sin \frac{1}{2}(\beta - \gamma) \sin \frac{1}{2}(\gamma - \alpha) \sin \frac{1}{2}(\alpha - \beta)$.

2. Prove that the locus of the poles of normal chords of the ellipse $\frac{x^2}{a^2} + \frac{y^2}{b^2} = 1$ is the curve $x^2y^2(a^2 - b^2)^2 = a^6y^2 + b^6x^2$.

3. Show that tangents to the ellipse $\frac{x^2}{a^2} + \frac{y^2}{b^2} = 1$ at the ends of a chord which subtend a right angle at the centre intersect on the ellipse $\frac{x^2}{a^4} + \frac{y^2}{b^4} = \frac{1}{a^2} + \frac{1}{b^2}$.

4. Straight lines are drawn through the foci of an ellipse perpendicular to a pair of conjugate diameters. Show that the locus of their point of intersection is an ellipse with eccentricity equal to that of the first ellipse.

5. A variable line meets an ellipse $\frac{x^2}{a^2} + \frac{y^2}{b^2} = 1$ at A and B. P is a fixed point on the ellipse. If PA and PB are parallel to conjugate diameters of the fixed ellipse $\frac{x^2}{a^2 + c^2} + \frac{y^2}{b^2 + c^2} = 1$, prove that AB passes through a fixed point on the normal at P to the first ellipse.

6. The pole of the normal at a point P on an ellipse is at Q. Prove that the product of the perpendiculars from Q and the centre of the ellipse to this normal is equal to the product of the focal distances from P.

7. P is the point θ on the ellipse $x = a \cos \theta$, $y = b \sin \theta$, P and F is the focus $(ae, 0)$. Show that the circle described on PF as diameter touches the auxiliary circle at the point $\left(\frac{a(e + \cos \theta)}{1 + e \cos \theta}, \frac{b \sin \theta}{1 + e \cos \theta}\right)$.

8. The normals at P and Q to a curve intersect in R. The limiting position of R as P and Q tend to coincidence is called the **centre of curvature** of the curve at P. Show that the centre of curvature at the point θ on the ellipse $x = a \cos \theta$, $y = b \sin \theta$ is the point $\left(\frac{(a^2 - b^2) \cos^3 \theta}{a}, -\frac{(a^2 - b^2) \sin^3 \theta}{b}\right)$.

9. The locus of the centre of curvature of a curve is called its **evolute.** Prove that the evolute of the ellipse $\frac{x^2}{a^2} + \frac{y^2}{b^2} = 1$ is the curve $(ax)^{2/3} + (by)^{2/3} = (a^2 - b^2)^{2/3}$.

10. C is the centre of curvature at the point P of a curve, the circle with centre C and radius CP is called the **circle of curvature** at P. Show that the equation of the circle of curvature at the point θ on the ellipse $x = a \cos \theta$, $y = b \sin \theta$ is

$$x^2 + y^2 - \frac{2(a^2 - b^2)x \cos^3 \theta}{a} + \frac{2(a^2 - b^2)y \sin^3 \theta}{b} - (2b^2 - a^2) \cos^2 \theta - (2a^2 - b^2) \sin^2 \theta = 0.$$

Answers

3. $(-nla^2/(l^2a^2 + m^2b^2), \; -mnb^2/(l^2a^2 + m^2b^2))$. **4.** $x_A(x - x_A)/a^2 + y_A(y - y_A)/b^2 = 0$.

CHAPTER V

HYPERBOLA

37. Hyperbola

In §29 the ellipse was defined as a curve with the geometrical property $PM^2 = k^2 A'M \,.\, MA$ (Fig. 20). We now examine the curve obtained by allowing $PM^2 = -k^2 A'M \,.\, MA$. Clearly $A'M$ and MA must have opposite signs if P is a real point. Formally, we define the hyperbola to be the locus of the point which moves so

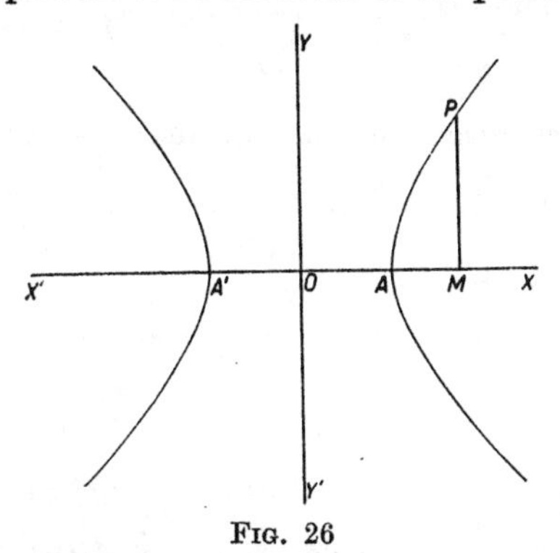

FIG. 26

that $PM^2 = -k^2 A'M \,.\, MA$, where A and A' are two fixed points (Fig. 26) and PM is perpendicular to $A'A$. Again let us choose the mid-point O of $A'A$ as origin and OA as the direction of the positive x-axis.

From the definition it follows that

$$y^2 = -k^2(x + a)(a - x) = -k^2(a^2 - x^2)$$

if $A'A = 2a$. Introduce $b^2 = k^2a^2$ and the equation of the hyperbola takes the standard form

$$S \equiv \frac{x^2}{a^2} - \frac{y^2}{b^2} - 1 = 0. \tag{37.1}$$

We refer to $A'A$ as the **transverse** axis, and its perpendicular bisector ($Y'OY$ in the figure) as the **conjugate** axis. A' and A are the **vertices** and O is the **centre,** bisecting all chords called **diameters** which pass through it. The transverse and conjugate axes are lines of symmetry. No real part of the curve exists for $-a < x < a$ and so the curve consists of two branches, extending to infinity in each of the four quadrants.

38. Tangent and polar properties

Many of the theorems developed for the ellipse also hold good for the hyperbola if we write $-b^2$ for b^2. With the notation

$$S_A \equiv \frac{x_A{}^2}{a^2} - \frac{y_A{}^2}{b^2} - 1; \quad T_A \equiv \frac{x_A x}{a^2} - \frac{y_A y}{b^2} - 1;$$

$$T_{AB} = \frac{x_A x_B}{a^2} - \frac{y_A y_B}{b^2} - 1;$$

the reader is advised to work out the detailed proofs of the following results:

(i) two points A and B are conjugate with respect to the hyperbola if $T_{AB} = 0$;

(ii) the polar of A is the straight line $T_A = 0$;

(iii) the pole of $lx + my + n = 0$ is at $(-a^2l/n, b^2m/n)$;

(iv) $l_1x + m_1y + n_1 = 0$ and $l_2x + m_2y + n_2 = 0$ are conjugate if $a^2l_1l_2 - b^2m_1m_2 = n_1n_2$;

(v) the tangent at A is the straight line $T_A = 0$;

(vi) the normal at A is $\frac{y_A}{b^2}(x - x_A) + \frac{x_A}{a^2}(y - y_A) = 0$;

(vii) $lx + my + n = 0$ touches the hyperbola if $a^2l^2 - b^2m^2 = n^2$;

(viii) the pair of tangents from A to the hyperbola is $S_AS - T_A{}^2 = 0$;

(ix) the polar of A is the line joining the points of contact of the tangents through A to the hyperbola;

(x) the orthoptic locus is the circle $x^2 + y^2 = a^2 - b^2$ and is accordingly real only if $a > b$.

Example 1. Show that the straight line $y = gx + \sqrt{a^2g^2 - b^2}$ touches the hyperbola $\frac{x^2}{a^2} - \frac{y^2}{b^2} = 1$ for all values of g.

39. Parametric equations

Draw the circle, Fig. 27, on $A'A$ as diameter. Construct the tangent MT to this circle, where M is the foot of the perpendicular from P to $A'A$. Denote the angle MOT by θ. Then $x_P = a \sec \theta$. Substitute in the equation of the hyperbola (37.1) and the result is $y_P = b \tan \theta$. As θ varies from 0 to $\pi/2$ the point P traverses that part of the hyperbola in the first quadrant starting from A and tending to infinity as θ tends to $\pi/2$. Next, as θ varies from $\pi/2$

to π, the point P traverses the portion of the hyperbola in the third† quadrant from infinity to A'. P then completes this branch out to infinity in the second quadrant as θ varies from π to $3\pi/2$. Finally, as θ varies from $3\pi/2$ to 2π, P returns to A from infinity in the fourth quadrant. Thus P covers every point of the hyperbola as θ ranges from 0 to 2π. Consequently, we may say that the hyperbola has the parametric equations

$$x = a \sec \theta; \quad y = b \tan \theta. \tag{39.1}$$

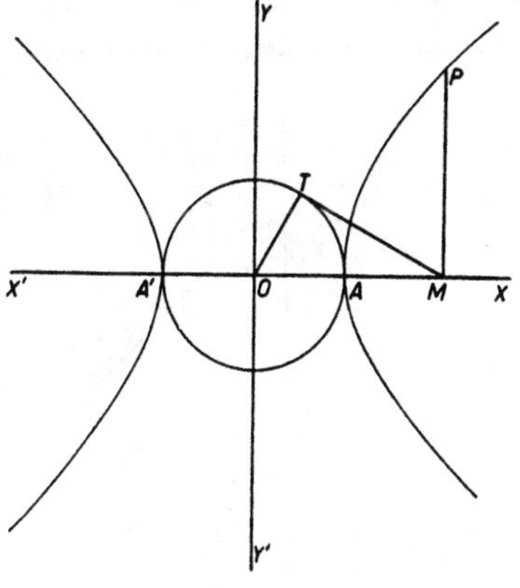

FIG. 27

The reader is asked to prove that the chord joining the two points α and β on the hyperbola has the equation

$$\frac{x}{a} \cos \tfrac{1}{2}(\alpha - \beta) - \frac{y}{b} \sin \tfrac{1}{2}(\alpha + \beta) = \cos \tfrac{1}{2}(\alpha + \beta). \tag{39.2}$$

Therefore, the tangent at θ is

$$\frac{x}{a} - \frac{y}{b} \sin \theta = \cos \theta. \tag{39.3}$$

Another parametric representation of the hyperbola is by the equations

$$x = a \cosh \theta; \quad y = b \sinh \theta, \tag{39.4}$$

which satisfy (37.1) because $\cosh^2 \theta - \sinh^2 \theta = 1$. There is a difficulty due to the fact that x is always positive for all real values

† We do not make $\pi/2 < \theta < \pi$ correspond to the second quadrant as this would imply $y_P = -b \tan \theta$ for this quadrant.

of θ and so the equations (39.4) represent only the right-hand branch of the hyperbola as θ ranges from $-\infty$ to $+\infty$.

The reader will verify that the equation of the chord joining the points α and β is

$$\frac{x}{a}\cosh\tfrac{1}{2}(\alpha+\beta)-\frac{y}{b}\sinh\tfrac{1}{2}(\alpha+\beta)=\cosh\tfrac{1}{2}(\alpha-\beta),$$

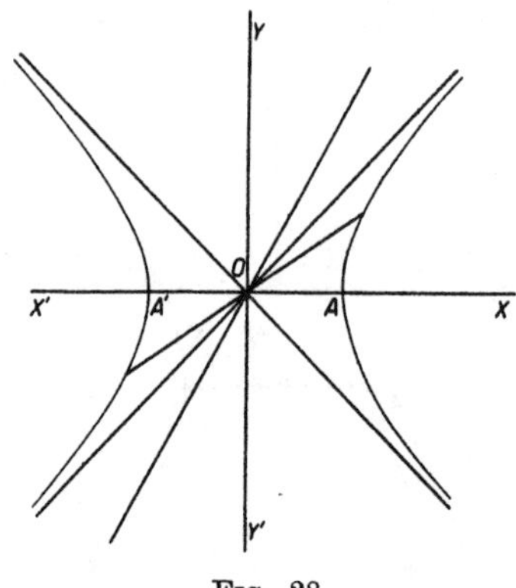

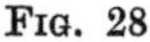

Fig. 28

whilst the tangent at the point θ is

$$\frac{x}{a}\cosh\theta-\frac{y}{b}\sinh\theta=1.$$

Example 2. Show that $lx+my+n=0$ is a normal to the hyperbola $\dfrac{x^2}{a^2}-\dfrac{y^2}{b^2}=1$ if $\dfrac{a^2}{l^2}-\dfrac{b^2}{m^2}=\dfrac{(a^2+b^2)^2}{n^2}$.

Example 3. Show that the tangents at the points α and β on the hyperbola $x=a\sec\theta$, $y=b\tan\theta$ intersect at $(a\cos\frac{1}{2}(\alpha-\beta)/\cos\frac{1}{2}(\alpha+\beta), b\sin\frac{1}{2}(\alpha+\beta)/\cos\frac{1}{2}(\alpha+\beta))$.

Example 4. Show that the tangents at the points α and β on the hyperbola $x=a\cosh\theta$, $y=b\sinh\theta$ intersect at $(a\cosh\frac{1}{2}(\alpha+\beta)/\cosh\frac{1}{2}(\alpha-\beta), b\sinh\frac{1}{2}(\alpha+\beta)/\cosh\frac{1}{2}(\alpha-\beta))$.

40. Asymptotes

The diameter $y=gx$ intersects the hyperbola $\dfrac{x^2}{a^2}-\dfrac{y^2}{b^2}=1$, at the two points where $x^2\left(\dfrac{1}{a^2}-\dfrac{g^2}{b^2}\right)=1$. These points are real if and only if $-b/a<g<b/a$. As g tends to $\pm b/a$ the points of intersection tend to infinity. Thus the two straight lines

$x/a + y/b = 0$ and $x/a - y/b = 0$, called the **asymptotes,** divide all diameters into two classes (Fig. 28). One class intersects the hyperbola but the other does not. From (39.3) the tangent at the point θ is $\dfrac{x}{a} - \dfrac{y}{b}\sin\theta = \cos\theta$ and so for $\theta = \pi/2$ we obtain $x/a - y/b = 0$ and for $\theta = 3\pi/2$ we have $x/a + y/b = 0$. Thus the asymptotes are the limiting positions of the tangent as the point of contact tends to infinity.

The angle between the asymptotes is $2\tan^{-1}(b/a)$.

Example 5. Show that a straight line parallel to an asymptote intersects a hyperbola in only one point.

Example 6. Prove that the tangent to a hyperbola (i) makes with the asymptotes a triangle of constant area, (ii) intercepted between two asymptotes is bisected at the point of contact.

Example 7. A straight line intersects a hyperbola at the points P, Q and its asymptotes at the points R and S. Prove that $PR = QS$.

41. Conjugate diameters

Let A and B be the points α and β on the hyperbola $x = a\sec\theta$, $y = b\tan\theta$. Then the co-ordinates of its mid-point P are

$$x_P = \tfrac{1}{2}a(\sec\alpha + \sec\beta) = \frac{a\cos\frac{1}{2}(\alpha+\beta)\cos\frac{1}{2}(\alpha-\beta)}{\cos\alpha\cos\beta},$$

$$y_P = \tfrac{1}{2}b(\tan\alpha + \tan\beta) = \frac{b\sin\frac{1}{2}(\alpha+\beta)\cos\frac{1}{2}(\alpha+\beta)}{\cos\alpha\cos\beta},$$

from which

$$y_P/x_P = b\sin\tfrac{1}{2}(\alpha+\beta)/a\cos\tfrac{1}{2}(\alpha-\beta).$$

The gradient of the chord AB from (39.2) is

$$g = b\cos\tfrac{1}{2}(\alpha-\beta)/a\sin\tfrac{1}{2}(\alpha+\beta).$$

Introduce $g' = b^2/a^2g$ and we have $y_P = g'x_P$. For a system of parallel chords AB, g is a constant. Therefore g' is constant. Thus the locus of the mid-points of all chords parallel to AB is the diameter OP (Fig. 29) with the equation $y = g'x$. Likewise, the locus of the mid-points of all chords parallel to OP is the diameter OQ parallel to AB. Hence conjugate diameters can also be defined for the hyperbola as diameters each of which bisects all chords parallel to the other.

The theory differs in several respects from that for the ellipse. The product of the gradients of conjugate diameters is given by

$$gg' = b^2/a^2.$$

It follows that one gradient is greater arithmetically than b/a, whilst the other is less arithmetically than b/a. Thus if a diameter intersects the hyperbola its conjugate does not. If $y = gx$ intersects the hyperbola $\frac{x^2}{a^2} - \frac{y^2}{b^2} = 1$, then the conjugate diameter $y = g'x$ will intersect the **conjugate hyperbola** $\frac{x^2}{a^2} - \frac{y^2}{b^2} = -1$, dotted in Fig. 29.

Note that each of the asymptotes is self-conjugate.

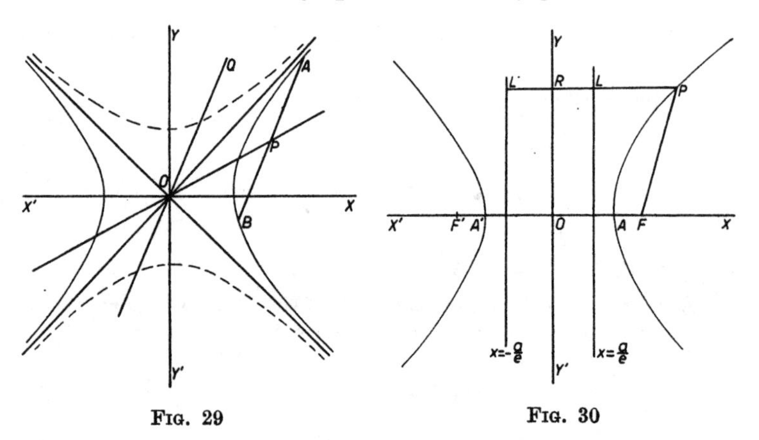

FIG. 29 FIG. 30

Example 8. Show that the tangents at the points where a diameter cuts a hyperbola and where the conjugate diameter cuts the conjugate hyperbola form a parallelogram with vertices lying on the common asymptotes.

42. Focus and directrix

Introduce the positive quantity e, called the **eccentricity** by the relation

$$e^2 = 1 + \frac{b^2}{a^2} \quad \text{or} \quad b^2 = a^2(e^2 - 1).$$

We also select the two points, called **foci**, $F \equiv (ae, 0)$ and $F' \equiv (-ae, 0)$ on the transverse axis of the hyperbola $\frac{x^2}{a^2} - \frac{y^2}{b^2} = 1$.

Since $e > 1$, F lies to the right of A and F' to the left of A' in Fig. 30. The polar of F, called the **directrix**, is the straight line $x = a/e$. If P is the point θ, we have

$$LP = RP - RL = \frac{a}{e}(e \sec \theta - 1),$$

$$PF^2 = (a \sec \theta - ae)^2 + b^2 \tan^2 \theta,$$
$$= a^2 \sec^2 \theta - 2a^2e \sec \theta + a^2e^2 + a^2(e^2 - 1) \tan^2 \theta,$$
$$= a^2(e \sec \theta - 1)^2.$$

Hence $$PF/LP = e.$$

Likewise, let PL cut the second directrix $x = -a/e$, which is the polar of the focus at $F' \equiv (-ae, 0)$, at L'. Then an easy calculation yields

$$L'P = \frac{a}{e}(e \sec \theta + 1),$$
$$PF'^2 = a^2(e \sec \theta + 1)^2,$$

and so $PF'/L'P = e$.

Thus, like the ellipse, the hyperbola can be defined as the locus of a point which moves so that its distance from a fixed point is in a constant ratio *greater* than unity to its distance from a fixed line.

Our calculations show us that $PF' - PF = 2a$. This is so if P lies on the right-hand branch. For the left-hand branch of the hyperbola we have $PF - PF' = 2a$.

Example 9. Show that the tangent at a point on a hyperbola is the internal bisector of the angle between the lines joining this point to the two foci.

43. Rectangular hyperbola

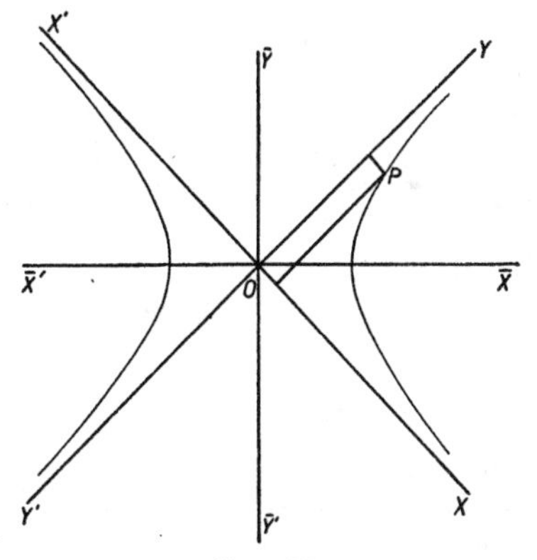

Fig. 31

The asymptotes of the hyperbola $\dfrac{x^2}{a^2} - \dfrac{y^2}{b^2} = 1$ are mutually orthogonal if $a = b$, and the curve is then called a **rectangular** hyperbola. We now proceed to simplify its equation by selecting the asymptotes as coordinate axes.

Label the old axes (Fig. 31) by $\bar{X}'O\bar{X}$, $\bar{Y}'O\bar{Y}$ and denote the coordinates referred to these axes by $\bar{x}$ and $\bar{y}$. Label the new axes along the asymptotes by XOX', YOY' and as usual reserve x and y for the coordinates referred to these axes. The equations of the asymptotes referred to the old axes are $\bar{x} + \bar{y} = 0$ and $\bar{x} - \bar{y} = 0$. Hence the perpendicular distances from $P \equiv (\bar{x}, \bar{y})$ to the asymptotes are $\frac{1}{\sqrt{2}}(\bar{x} + \bar{y})$ and $\frac{1}{\sqrt{2}}(\bar{x} - \bar{y})$. These perpendicular distances are the coordinates x and y referred to the new axes. Thus

$$x = \frac{1}{\sqrt{2}}(\bar{x} - \bar{y}); \quad y = \frac{1}{\sqrt{2}}(\bar{x} + \bar{y}).$$

The equation $\bar{x}^2 - \bar{y}^2 = a^2$ of the rectangular hyperbola then becomes

$$xy = c^2$$

if we introduce $c^2 = \frac{1}{2}a^2$.

Parametrically the rectangular hyperbola is represented by the equations

$$x = ct; \quad y = \frac{c}{t}.$$

Example 10. For the rectangular hyperbola $x = ct$, $y = c/t$, show that

(i) the chord joining the points t_1 and t_2 is $x + t_1t_2y - c(t_1 + t_2) = 0$,
(ii) the tangent at t is $x + t^2y - 2ct = 0$,
(iii) the normal at t is $t^3x - ty + c(1 - t^4) = 0$,
(iv) the normals at the four points t_1, t_2, t_3 and t_4 are concurrent if $t_1t_2t_3t_4 = -1$ and $\Sigma t_1t_2 = 0$,
(v) the normals at the three points t_1, t_2 and t_3 are concurrent if $t_1t_2t_3(t_2t_3 + t_3t_1 + t_1t_2) = t_1 + t_2 + t_3$,
(vi) the points t_1, t_2, t_3 and t_4 are concyclic if $t_1t_2t_3t_4 = 1$.

Example 11. P, Q, R and S are four points on a rectangular hyperbola such that PQ is perpendicular to RS. Show that the product of the gradients of OP, OQ, OR and OS is unity, where O is the centre of the hyperbola.

Miscellaneous examples

1. Show that the necessary and sufficient conditions that the normals at the four points θ_1, θ_2, θ_3 and θ_4 on the hyperbola $x = a \sec \theta$, $y = b \tan \theta$ be concurrent are $t_1t_2t_3t_4 = -1$ and $\Sigma t_1t_2 = 0$, where $t = \tan \theta/2$. If the parametric equations are $x = a \cosh \theta$, $y = b \sinh \theta$ the results are the same but now $t = \tanh \theta/2$.

2. Show that the necessary and sufficient condition that the four points θ_1, θ_2, θ_3 and θ_4 on the hyperbola $x = a \sec \theta$, $y = b \tan \theta$ be concyclic is $\Sigma t_1t_2t_3 = -\Sigma t_1$, where $t = \tan \theta/2$. If the parametric equations are $x = a \cosh \theta$, $y = b \sinh \theta$ the results are the same but $t = \tanh \theta/2$.

3. A circle cuts a rectangular hyperbola in four points. Show that the centroid of these four points bisects the join of the centres of the circle and the hyperbola.

Further prove that the sum of the squares of the distances of these four points from the centre of the hyperbola is equal to the square on the diameter of the circle.

4. The straight line $lx + my + n = 0$ intersects the hyperbola $\frac{x^2}{a^2} - \frac{y^2}{b^2} = 1$ at the points L and M. Prove that the equation of the circle on LM as diameter is $(a^2l^2 - b^2m^2)(x^2 + y^2) + 2n(a^2lx - b^2my) + a^2b^2(l^2 + m^2) + n^2(a^2 - b^2) = 0$.

5. Show that the orthocentre of a triangle inscribed in a rectangular hyperbola lies on the hyperbola.

6. The tangents at the points A, A' on a hyperbola cut an asymptote at B, B' respectively. Show that AA' bisects BB'.

7. From the point P on a rectangular hyperbola, the three normals PA, PB and PC are drawn to the hyperbola. Prove that the centroid of the triangle ABC is the centre of the hyperbola.

8. The normal at P to a hyperbola of eccentricity e intersects the transverse and conjugate axes at L and M. Show that the locus of the mid-point of LM is a hyperbola of eccentricity $e/\sqrt{e^2 - 1}$.

9. Show that the centre of curvature (see Miscellaneous Examples 8 and 9, Chapter IV) of the hyperbola $x = a \sec\theta$, $y = b \tan\theta$ at the point θ is $\left(\frac{(a^2 + b^2)\sec^3\theta}{\theta}, -\frac{(a^2 + b^2)\tan^3\theta}{\theta}\right)$. Further show that the evolute of the hyperbola is the curve $(ax)^{2/3} - (by)^{2/3} = (a^2 + b^2)^{2/3}$.

10. Prove that the line joining the mid-points of the diagonals of a quadrilateral which circumscribes a hyperbola (or ellipse) passes through its centre.

11. P is a point on, and O is the centre of, the hyperbola $\frac{x^2}{a^2} - \frac{y^2}{b^2} = 1$. The diameter conjugate to OP intersects the conjugate hyperbola $\frac{x^2}{a^2} - \frac{y^2}{b^2} = -1$ at R. Show that the locus of the point of intersection of the normals at P and R is the line-pair $a^2x^2 - b^2y^2 = 0$.

CHAPTER VI

PARABOLA

44. Parabola

We have seen that the eccentricity of an ellipse is less than unity whilst that of a hyperbola is greater than unity. This suggests that we investigate the intermediate case of a curve with unit eccentricity. Formally a **parabola** is the locus of a point P which moves so that its distance from a fixed point, called the **focus**, is equal to its distance from a fixed line called the **directrix.**

In Fig. 32, let F be the focus and $R'R$ the directrix. Draw FD perpendicular to the directrix and select the mid-point O of FD as

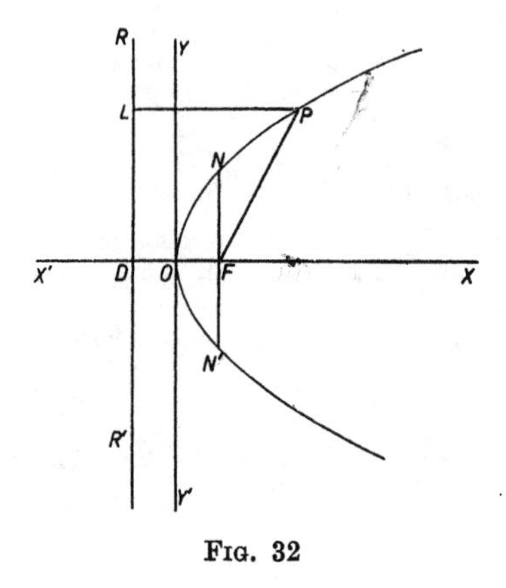

FIG. 32

origin, and OF as the direction of the positive x-axis. Let $OF = a$, then $PF = PL$ yields the equation

$$(x - a)^2 + y^2 = (x + a)^2,$$

which reduces to the standard form

$$S \equiv y^2 - 4ax = 0. \tag{44.1}$$

The focus is at $(a, 0)$ and the equation of the directrix is $x = -a$. O is the **vertex** and OX is the **axis.** The chord $N'N$ through the focus perpendicular to the axis is called the **latus-rectum** and it is of length $4a$. If (x, y) lies on the parabola, so does $(x, -y)$. Hence the parabola is symmetrical about its axis.

Example 1. Find the equation of the parabola whose focus is at $(-2, 1)$ and whose directrix is $x - 2y + 3 = 0$. See Loney, P176, Section 199

45. Tangent and polar properties

Again we may repeat the theory centred around Joachimsthal's quadratic equation. With the notation

$$S_A \equiv y_A{}^2 - 4ax_A; \quad T_A = y_A y - 2a(x + x_A);$$
$$T_{AB} = y_A y_B - 2a(x_A + x_B);$$

the reader may verify the following results:

(i) the two points A and B are conjugate if $T_{AB} = 0$;

(ii) the polar of A is the straight line $T_A = 0$;

(iii) the pole of $lx + my + n = 0$ is at $(n/l, -2am/l)$;

(iv) $l_1x + m_1y + n_1 = 0$ and $l_2x + m_2y + n_2 = 0$ are conjugate if $2am_1m_2 - n_1l_2 - n_2l_1 = 0$;

(v) the tangent at A is $T_A = 0$;

(vi) the normal at A is $2a(y - y_A) + y_A(x - x_A) = 0$;

(vii) $lx + my + n = 0$ touches the parabola if $am^2 - nl = 0$;

(viii) the pair of tangents from A to the parabola is $S_A S - T_A{}^2 = 0$;

(ix) the polar of A is the line joining the points of contact of the tangents through A to the parabola;

(x) the orthoptic locus is the directrix $x + a = 0$. It is no longer a circle as was the case with the ellipse and the hyperbola.

Example 2. Verify that the directrix of the parabola is the polar of its focus.

Example 3. Show that the straight line $y = gx + a/g$ touches the parabola $y^2 = 4ax$ for all values of g.

46. Parametric equations

The elimination of t from

$$x = at^2; \quad y = 2at \tag{46.1}$$

yields $y^2 = 4ax$. Thus $x = at^2$; $y = 2at$ are parametric equations of the parabola.

The equation of the chord joining the points t_1 and t_2 is

$$\frac{x - at_1{}^2}{at_1{}^2 - at_2{}^2} = \frac{y - 2at_1}{2at_1 - 2at_2},$$

which reduces to

$$2x - (t_1 + t_2)y + 2at_1t_2 = 0. \tag{46.2}$$

Putting $t_1 = t_2 = t$, the equation of the tangent at the point t is

$$x - ty + at^2 = 0. \tag{46.3}$$

Consequently the normal at the point t is $t(x - at^2) + y - 2at = 0$. That is,

$$tx + y - at^3 - 2at = 0. \tag{46.4}$$

Example 4. Prove that the line joining the points t_1 and t_2 passes through the focus if $t_1t_2 = -1$. Hence deduce that the tangents at the ends of a focal chord are mutually orthogonal.

Example 5. Show that the locus of the foot of the perpendicular from the focus to a tangent of a parabola is the tangent at the vertex.

Example 6. Show that the tangents at the points t_1 and t_2 on the parabola $(at^2,\ 2at)$ intersect at $(at_1t_2,\ a(t_1 + t_2))$ and the normals intersect at $(a(t_1^2 + t_1t_2 + t_2^2 + 2),\ -at_1t_2(t_1 + t_2))$.

47. Diameters

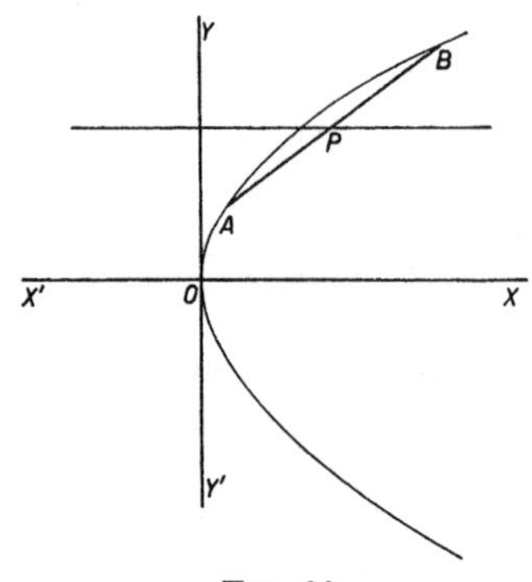

FIG. 33

Let us find the locus of the mid-points of the system of parallel chords AB with gradient g. In Fig. 33 let $A \equiv (at_1^2,\ 2at_1)$ and $B \equiv (at_2^2,\ 2at_2)$. The gradient of AB is $2a(t_1 - t_2)/a(t_1^2 - t_2^2)$ $= 2/(t_1 + t_2) = g$, whilst the coordinates of P are

$$x_P = \tfrac{1}{2}a(t_1^2 + t_2^2); \quad y_P = a(t_1 + t_2) = \frac{2a}{g}.$$

Hence the locus of P is the straight line $y = 2a/g$ parallel to the x-axis. In the case of the parabola we use the term **diameter** for any straight line parallel to the axis and we say that the direction with gradient g is conjugate to the diameter $y = 2a/g$.

Example 7. Show that a diameter intersects a parabola at the point of contact of the conjugate tangent line.

48. Normals

From (46.4) the equation of the normal at the point $(at^2, 2at)$ on the parabola $y^2 = 4ax$ is

$$tx + y - at^3 - 2at = 0.$$

Let A be any fixed point on this normal. Then

$$tx_A + y_A - at^3 - 2at = 0.$$

This is a cubic equation and so three normals can in general be drawn from a point to a parabola. There is no term in t^2 in this equation and so we deduce the necessary and sufficient condition that the normals at the three points t_1, t_2 and t_3 be concurrent is $t_1 + t_2 + t_3 = 0$.

Example 8. Find the condition that the normals at the points t_1 and t_2 on the parabola $x = at^2$, $y = 2at$ be orthogonal. Hence obtain the locus of the points of intersection of orthogonal normals to the parabola.

Miscellaneous examples

1. Show that the four points with parameters t_1, t_2, t_3 and t_4 on the parabola $y^2 = 4ax$ are conclyclic if $t_1 + t_2 + t_3 + t_4 = 0$.

2. Obtain the locus of the points of intersection of two tangents to a parabola $y^2 = 4ax$ which make an angle θ with one another.

3. Find the locus of the mid-points of chords of the parabola $y^2 = 4ax$ which subtend a right angle at the vertex.

4. Prove that the radical axis of the two circles, described on any two focal chords as diameters, passes through the vertex of the parabola.

5. Show that the centre of curvature (see Examples 8 and 9, page 44) of the parabola $x = at^2$, $y = 2at$ at the point t is $(2a + 3at^2, -2at^3)$. Further show that the evolute of the parabola is the curve $27ay^2 = 4(x - 2a)^3$.

6. Prove that the orthocentre of the triangle formed by three tangents to a parabola lies on the directrix.

7. Show that the area of the triangle formed by the points t_1, t_2 and t_3 on the parabola $x = at^2$, $y = 2at$ is $a^2(t_2 - t_3)(t_3 - t_1)(t_1 - t_2)$. Further the area of the triangle formed by the tangents at these points has half this value.

8. The normals to the parabola at the points A, B and C are concurrent. Show that the centroid of the triangle ABC is on the axis of the parabola.

9. Prove that an infinite number of triangles can be inscribed in the parabola $y^2 = 4ax$ whose sides touch the parabola $x^2 = 4by$. Further show that the circumcircles of all such triangles pass through the common vertex.

10. From any point P on a parabola $y^2 = 4ax$, straight lines PQ and PR are drawn normal to the parabola at Q and R respectively. Show that QR passes through the point $(-2a, 0)$.

Answers

1. $4x^2 + 4xy + y^2 + 14x + 2y + 16 = 0$. **8.** $y^2 = a(x - 3a)$.

Miscellaneous examples. 2. $y^2 - 4ax - (x + a)^2 \tan^2 \theta = 0$. **3.** $y^2 - 2ax + 8a^2 = 0$.

CHAPTER VII

TRANSFORMATION OF AXES

49. Translation of axes

We have investigated the ellipse, hyperbola and parabola with respect to the special choices of coordinate axes which yield the so-called standard forms $\frac{x^2}{a^2} + \frac{y^2}{b^2} = 1$, $\frac{x^2}{a^2} - \frac{y^2}{b^2} = 1$, $xy = c^2$ and $y^2 = 4ax$. Occasionally it may be necessary to employ some other pair of mutually orthogonal straight lines as coordinate axes. A change of axes is called a **transformation** of axes. We begin with

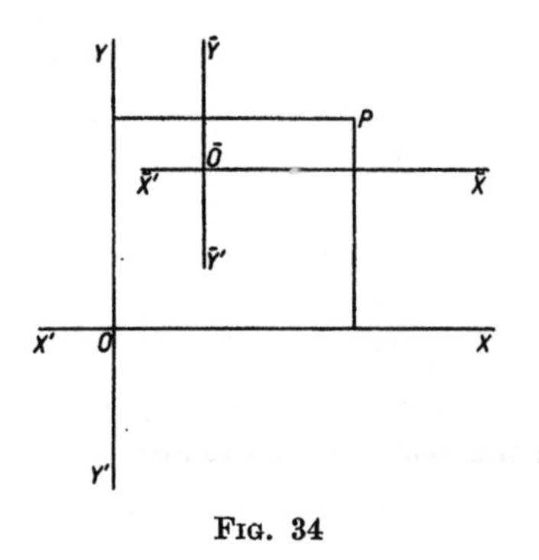

FIG. 34

the discussion of the simple case where there is no change in the direction of the axes, but there is a shift of origin. This we refer to as a **translation** of axes.

Let the point P (Fig. 34) have coordinates (x, y) referred to the OX, OY coordinate system (called the *old* coordinate system) and coordinates $(\bar{x}, \bar{y})$ referred to a *new* coordinate system $\bar{O}\bar{X}$, $\bar{O}\bar{Y}$ obtained by translating the origin from O to $\bar{O}$. Further, let the coordinates of the new origin $\bar{O}$ be (α, β) referred to the old axes OX, OY. From the figure it is clear that

$$x = \bar{x} + \alpha; \quad y = \bar{y} + \beta. \tag{49.1}$$

The change back again from the new axes to the old axes is called the **inverse** transformation and we have

$$\bar{x} = x - \alpha; \quad \bar{y} = y - \beta.$$

If $f(x, y) = 0$ represents a curve with respect to the old axes, then we see that $f(\bar{x} + \alpha, \bar{y} + \beta) = 0$ is the equation referred to the new axes. On the other hand, if $f(\bar{x}, \bar{y}) = 0$ is the equation of the curve with respect to the new axes, then $f(x - \alpha, y - \beta) = 0$ is the equation referred to the old axes.

Example 1. Translate to parallel axes through $(4, -2)$ and hence show that $16x^2 + 25y^2 - 128x + 100y - 44 = 0$ represents an ellipse with eccentricity $3/5$.

50. Invariants of a quadratic form under translation

We call $ax^2 + 2hxy + by^2 + 2gx + 2fy + c$ a **quadratic** form and ask: "What functions of a, b, c, f, g and h are unaltered by a translation of axes?" These functions are called **invariants** of the quadratic form. We note carefully that invariants are not functions of either x or y.

The translation (49.1) transforms the quadratic form

$$S \equiv ax^2 + 2hxy + by^2 + 2gx + 2fy + c$$

into

$$\bar{S} \equiv a(\bar{x} + \alpha)^2 + 2h(\bar{x} + \alpha)(\bar{y} + \beta) + b(\bar{y} + \beta)^2 + 2g(\bar{x} + \alpha) + 2f(\bar{y} + \beta) + c.$$

Let us write

$$\bar{S} \equiv \bar{a}\bar{x}^2 + 2\bar{h}\bar{x}\bar{y} + \bar{b}\bar{y}^2 + 2\bar{g}\bar{x} + 2\bar{f}\bar{y} + \bar{c}.$$

On equating coefficients, the relations connecting the *old* coefficients a, b, c, f, g, h and the *new* coefficients are

$$\left.\begin{aligned} &\bar{a} = a; \qquad \bar{h} = h; \qquad \bar{b} = b; \\ &\bar{g} = a\alpha + h\beta + g; \qquad \bar{f} = h\alpha + b\beta + f; \\ &\bar{c} = a\alpha^2 + 2h\alpha\beta + b\beta^2 + 2g\alpha + 2f\beta + c. \end{aligned}\right\} \qquad (50.1)$$

It is immediately obvious that a, b and h are invariants.

If $S = 0$ represents a line-pair so does $\bar{S} = 0$. The respective conditions are $\Delta = 0$ and $\bar{\Delta} = 0$, where

$$\Delta \equiv abc + 2fgh - af^2 - bg^2 - ch^2 \equiv \begin{vmatrix} a & h & g \\ h & b & f \\ g & f & c \end{vmatrix},$$

$$\bar{\Delta} = \bar{a}\bar{b}\bar{c} + 2\bar{f}\bar{g}\bar{h} - \bar{a}\bar{f}^2 - \bar{b}\bar{g}^2 - \bar{c}\bar{h}^2 \equiv \begin{vmatrix} \bar{a} & \bar{h} & \bar{g} \\ \bar{h} & \bar{b} & \bar{f} \\ \bar{g} & \bar{f} & \bar{c} \end{vmatrix}.$$

Thus there must be a relation between Δ and $\bar{\Delta}$ such that the vanishing of one implies the vanishing of the other. In fact, a simple calculation with the aid of (50.1) yields† $\Delta = \bar{\Delta}$. Accordingly Δ is an invariant.

In effect, we have eliminated α and β from the six equations (50.1) and so we do not expect more than four independent invariants. Consequently all the invariants of the quadratic form are a, b, h, Δ and any function of these four quantities.

51. Rotation of axes

A change of axes in which the origin remains fixed is called a **rotation** of axes. Let the point P (Fig. 35) have coordinates (x, y)

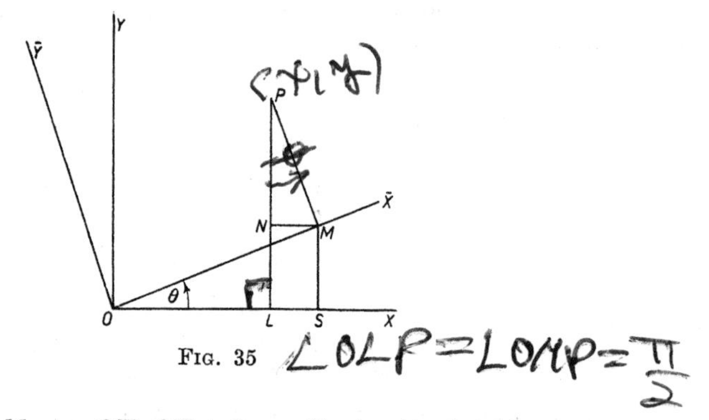

FIG. 35

referred to the *old* axes OX, OY and coordinates $(\bar{x}, \bar{y})$ with respect to a *new* coordinate system obtained by rotating OX to $O\bar{X}$ in a counter-clockwise direction through an angle $XO\bar{X} = \theta$. In the figure draw PL and PM perpendicular to OX and $O\bar{X}$ respectively and MN, MS perpendicular and parallel to PL respectively. O, L, M and P are concyclic points because the angle $OLP =$ the angle $OMP = \pi/2$. Hence the angle $LPM =$ the angle $LOM = \theta$. We have $OL = OS - NM$ and $LP = SM + NP$. Therefore,

$$\left.\begin{aligned} x &= \bar{x}\cos\theta - \bar{y}\sin\theta, \\ y &= \bar{x}\sin\theta + \bar{y}\cos\theta. \end{aligned}\right\} \qquad (51.1)$$

† The calculation is simple if we use determinant theory. We have

$$\bar{\Delta} = \begin{vmatrix} \bar{a} & \bar{h} & \bar{g} \\ \bar{h} & \bar{b} & \bar{f} \\ \bar{g} & \bar{f} & \bar{c} \end{vmatrix} = \begin{vmatrix} a & h & a\alpha + h\beta + g \\ h & b & h\alpha + b\beta + f \\ a\alpha + h\beta + g & h\alpha + b\beta + f & \begin{matrix} a\alpha^2 + 2h\alpha\beta + b\beta^2 \\ + 2g\alpha + 2f\beta + c \end{matrix} \end{vmatrix}$$

Subtract α times the first column and β times the second column from the third column. Next, subtract α times the first row and β times the second row from the third row. The resulting determinant is Δ.

The inverse transformation is obtained by rotating the axes through an angle θ in the clockwise direction and so we can interchange the pair of coordinates (x, y) with the pair $(\bar{x}, \bar{y})$ if we change θ into $-\theta$. The result is

$$\bar{x} = x \cos\theta + y \sin\theta,$$
$$\bar{y} = -x \sin\theta + y \cos\theta.$$

These equations also follow as the solution of the simultaneous equations (51.1) in $\bar{x}$ and $\bar{y}$.

If $f(x, y) = 0$ represents a curve with respect to the old axes, we see that $f(\bar{x}\cos\theta - \bar{y}\sin\theta, \bar{x}\sin\theta + \bar{y}\cos\theta) = 0$ is its equation referred to the new axes.

Example 2. Write down the equations for a rotation of axes through an angle $\pi/6$. Hence, show that the curve $\sqrt{3}x^2 - 2xy - \sqrt{3}y^2 + 4 = 0$ is a rectangular hyperbola.

Example 3. Write down the equations for a rotation of axes through an angle $\pi/4$. Hence, verify that the equation $2xy = a^2$ represents the rectangular hyperbola $x^2 - y^2 = a^2$ referred to its asymptotes.

52. Invariants of a quadratic form under rotation

The rotation (51.1) transforms the quadratic form

$$S \equiv ax^2 + 2hxy + by^2 + 2gx + 2fy + c$$

into

$$\begin{aligned} \bar{S} \equiv a(\bar{x}\cos\theta - \bar{y}\sin\theta)^2 &+ 2h(\bar{x}\cos\theta - \bar{y}\sin\theta)(\bar{x}\sin\theta + \bar{y}\cos\theta) \\ &+ b(\bar{x}\sin\theta + \bar{y}\cos\theta)^2 + 2g(\bar{x}\cos\theta - \bar{y}\sin\theta) \\ &+ 2f(\bar{x}\sin\theta + \bar{y}\cos\theta) + c. \end{aligned}$$

Again, let us write

$$\bar{S} \equiv \bar{a}\bar{x}^2 + 2\bar{h}\bar{x}\bar{y} + \bar{b}\bar{y}^2 + 2\bar{g}\bar{x} + 2\bar{f}\bar{y} + \bar{c},$$

and the relations obtained by equating coefficients are

$$\left.\begin{aligned} \bar{a} &= a\cos^2\theta + 2h\sin\theta\cos\theta + b\sin^2\theta, \\ \bar{h} &= (b - a)\sin\theta\cos\theta + h(\cos^2\theta - \sin^2\theta), \\ \bar{b} &= a\sin^2\theta - 2h\sin\theta\cos\theta + b\cos^2\theta, \\ \bar{g} &= g\cos\theta + f\sin\theta, \\ \bar{f} &= -g\sin\theta + f\cos\theta, \\ \bar{c} &= c. \end{aligned}\right\} \quad (52.1)$$

It is easy to verify that $\bar{a} + \bar{b} = a + b$ and $\bar{g}^2 + \bar{f}^2 = g^2 + f^2$, and so we obtain three invariants $a + b$, $g^2 + f^2$ and c. Now (52.1) comprise six equations and the elimination of θ suggests that there are five independent invariants. We suspect that Δ is again an invariant but the direction calculation is laborious and we provide the following geometrical argument:

In virtue of (51.1) we have $x^2 + y^2 = \bar{x}^2 + \bar{y}^2$. In fact, each expression measures the square of the distance from the fixed origin. Thus for any given value of λ the quadratic form

$$S_\lambda \equiv ax^2 + 2hxy + by^2 + 2gx + 2fy + c + \lambda(x^2 + y^2 + 1)$$

transforms by rotation of axes into

$$\bar{S}_\lambda \equiv \bar{a}\bar{x}^2 + 2\bar{h}\bar{x}\bar{y} + \bar{b}\bar{y}^2 + 2\bar{g}\bar{x} + 2\bar{f}\bar{y} + \bar{c} + \lambda(\bar{x}^2 + \bar{y}^2 + 1).$$

$S_\lambda = 0$ represents a line-pair if $\begin{vmatrix} a+\lambda & h & g \\ h & b+\lambda & f \\ g & f & c+\lambda \end{vmatrix} = 0.$

That is, λ is a root of the cubic equation

$$\lambda^3 + (a + b + c)\lambda^2 + (bc + ca + ab - f^2 - g^2 - h^2)\lambda + \Delta = 0.$$

Similarly $\bar{S}_\lambda = 0$ represents a line-pair if

$$\lambda^3 + (\bar{a} + \bar{b} + \bar{c})\lambda^2 + (\bar{b}\bar{c} + \bar{c}\bar{a} + \bar{a}\bar{b} - \bar{f}^2 - \bar{g}^2 - \bar{h}^2)\lambda + \bar{\Delta} = 0.$$

But if $S_\lambda = 0$ is a line-pair for some value of λ, then $\bar{S}_\lambda = 0$ must be a line-pair for the *same* value of λ because $\bar{S}_\lambda = 0$ and $S_\lambda = 0$ represent the same curve with respect to different coordinate axes. Hence the cubic equations in λ are identical and we have

$$a + b + c = \bar{a} + \bar{b} + \bar{c},$$
$$bc + ca + ab - f^2 - g^2 - h^2 = \bar{b}\bar{c} + \bar{c}\bar{a} + \bar{a}\bar{b} - \bar{f}^2 - \bar{g}^2 - \bar{h}^2,$$
$$\Delta = \bar{\Delta}.$$

We have already shown that $a + b$, $g^2 + f^2$ and c are invariants; the middle equation can be written

$$c(a + b) - (g^2 + f^2) + ab - h^2 = \bar{c}(\bar{a} + \bar{b}) - (\bar{g}^2 + \bar{f}^2) + \bar{a}\bar{b} - \bar{h}^2$$

and yields $ab - h^2 = \bar{a}\bar{b} - \bar{h}^2$.

Thus all the invariants of a quadratic form under rotation are $a + b$, c, $g^2 + f^2$, $ab - h^2$, Δ and any function of these five quantities.

53. General transformation of axes

A pair of orthogonal straight lines can be moved into coincidence with another orthogonal pair in the same plane by means of either (i) a translation followed by a rotation or (ii) a rotation followed by a translation.

(i) In Fig. 36, the axes OX, OY are translated to the parallel axes $\bar{O}X'$, $\bar{O}Y'$, and these are rotated about $\bar{O}$ to take up the position $\bar{O}\bar{X}$, $\bar{O}\bar{Y}$. Let us use (x', y') as coordinates referred to $\bar{O}X'$, $\bar{O}Y'$, then we have

$$x = x' + \alpha; \quad x' = \bar{x} \cos \theta - \bar{y} \sin \theta;$$
$$y = y' + \beta; \quad y' = \bar{x} \sin \theta + \bar{y} \cos \theta,$$

where (α, β) are the coordinates of $\bar{O}$ with respect to OX, OY.

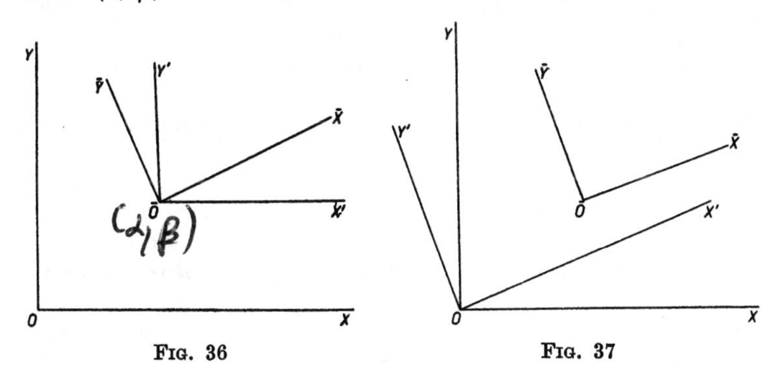

FIG. 36 FIG. 37

Consequently

$$\left.\begin{aligned} x &= \bar{x} \cos \theta - \bar{y} \sin \theta + \alpha, \\ y &= \bar{x} \sin \theta + \bar{y} \cos \theta + \beta, \end{aligned}\right\} \quad (53.1)$$

are the equations for the transformation of axes OX, OY to $\bar{O}\bar{X}$, $\bar{O}\bar{Y}$,

(ii) Let us now rotate (Fig. 37) from OX, OY to OX', OY' parallel. respectively, to $\bar{O}\bar{X}$, $\bar{O}\bar{Y}$ and then translate to parallel axes through $\bar{O}$. We have

$$x = x' \cos \theta - y' \sin \theta; \quad x' = \bar{x} + \alpha';$$
$$y = x' \sin \theta + y' \cos \theta; \quad y' = \bar{y} + \beta';$$

where (α', β') are the coordinates of $\bar{O}$ referred to OX', OY'. That is

$$\alpha' = \alpha \cos \theta + \beta \sin \theta; \quad \beta' = -\alpha \sin \theta + \beta \cos \theta.$$

Thus

$$x = (\bar{x} + \alpha \cos \theta + \beta \sin \theta) \cos \theta - (\bar{y} - \alpha \sin \theta + \beta \cos \theta) \sin \theta,$$
$$y = (\bar{x} + \alpha \cos \theta + \beta \sin \theta) \sin \theta + (\bar{y} - \alpha \sin \theta + \beta \cos \theta) \sin \theta,$$

which simplify to (53.1).

The method of (ii) is not so simple as (i) and so we prefer in an example to carry out a general transformation of axes by first translating the axes and then performing the rotation.

The general transformation involves three quantities α, β, θ. Accordingly we now expect to obtain only three invariants. Under a translation the invariants are a, b, h and Δ, whilst under a rotation the invariants are $a + b$, $g^2 + f^2$, c, $ab - h^2$ and Δ. Consequently the invariants under a general transformation are $a + b$, $ab - h^2$ and Δ. We introduce the notation

$$I \equiv a + b; \quad J \equiv ab - h^2.$$

54. Absolute invariants

The invariants already obtained are with respect to the quadratic form $S \equiv ax^2 + 2hxy + by^2 + 2gx + 2fy + c$. We are usually more interested in the curve $S = 0$ than the quadratic form S. The curve $\rho S = 0$, where ρ is any numerical factor, is identical with the curve $S = 0$ but the quadratic form ρS is not the same as the quadratic form S and their invariants are therefore different. If we agree in a general transformation of axes not to multiply by any extraneous number ρ, then we may say that I, J and Δ are invariants of the curve $S = 0$. On that account we often refer to them as **relative invariants.**

On the other hand, we may be told that $S = 0$ and $\bar{S} = 0$ represent the same curve with respect to different sets of axes. But $\bar{S}$ may be a numerical factor say ρ times the transformed quadratic form corresponding to S. Then the invariants of $\bar{S}$ which we shall call $\bar{I}$, $\bar{J}$ and $\bar{\Delta}$ are connected with the invariants of S by the obvious relations

$$\bar{I} = \rho I; \quad \bar{J} = \rho^2 J; \quad \bar{\Delta} = \rho^3 \Delta.$$

Thus we can form two independent invariants by the elimination of ρ. These are called **absolute invariants** and we may choose any two of the following: I^2/J, Δ/I^3, Δ^2/J^3, $\Delta I/J^2$, etc. We shall see that any geometrical feature independent of the choice of axes can be expressed in terms of absolute invariants.

Miscellaneous examples

1. Transform the following equations according to the instructions indicated:

(*a*) $x^2 - 4xy + y^2 + 8x + 2y - 5 = 0$ to new axes through $(2, 3)$ rotated through an angle $45°$.

(*b*) $3x^2 + 8xy - 3y^2 + 6x + 8y + 4 = 0$ to new axes through $(-1, 0)$ rotated through an angle $\tan^{-1} \frac{1}{2}$.

(*c*) $x^2 - 6xy + 9y^2 + 4x + 8y + 15 = 0$ to new axes through $(-2, -1)$ rotated through an angle $\tan^{-1} \frac{1}{3}$.

(*d*) $3x^2 - 24xy + 10y^2 + 6x + 52y = 0$ to new axes through $(3, 1)$ rotated through an angle $\tan^{-1} \frac{3}{4}$.

2. Calculate the angle subtended by the chord $2x + 3y = 1$ of the rectangular hyperbola $x^2 - y^2 = 1$ at the point (1, 2).

3. Obtain λ in terms of θ if the curves $x^2 \cos^2 \theta + 2\lambda xy \cos \theta \sin \theta + y^2 \sin^2 \theta = 1$ and $x^2 \sec^2 \theta - y^2 \tan^2 \theta = 1$ are identical.

4. Calculate λ and μ if the two curves $2x^2 - 3y^2 = 7$ and $x^2 + 10xy + \lambda y^2 - 14x - 22y + 3\mu = 0$ are identical.

Answers

Miscellaneous examples. 1. (*a*) $x^2 - 3y^2 = 6$, (*b*) $5(x^2 - y^2) = -1$, (*c*) $\sqrt{10}y^2 + 2x = 0$, (*d*) $6x^2 - 19y^2 = 35$. **2.** $\tan^{-1}\{\sqrt{294}/2\}$. **3.** $\pm\sqrt{1 + \sec^6 \theta}$. **4.** Either $\lambda = 1$, $\mu = 13$ or $\lambda = -19/6$, $\mu = 13571/3042$.

CHAPTER VIII

GENERAL EQUATION OF THE SECOND DEGREE

55. General quadratic equation

We saw that the equation of a line-pair, circle, ellipse, hyperbola and parabola are all of the second degree. Now we shall demonstrate the converse that every equation of the second degree represents one of the above curves.

Consider the curve

$$S \equiv ax^2 + 2hxy + by^2 + 2gx + 2fy + c = 0, \qquad (55.1)$$

which becomes

$$S' \equiv a'x'^2 + 2h'x'y' + b'y'^2 + 2g'x' + 2f'y' + c' = 0,$$

if we rotate axes through an angle θ in the counter-clockwise direction. From (52.1) we have

$$\begin{aligned} h' &= (b - a) \sin \theta \cos \theta + h(\cos^2 \theta - \sin^2 \theta), \\ &= \tfrac{1}{2}(b - a) \sin 2\theta + h \cos 2\theta. \end{aligned}$$

For all values of a, b and h ($a = b = h = 0$ is excluded as then S is not quadratic) we can choose the angle $\theta = \frac{1}{2} \tan^{-1} \{2h/(a - b)\}$. This ensures that $h' = 0$. That is, a rotation of axes has been found which removes the product term $x'y'$. The curve is now represented by

$$S' \equiv a'x'^2 + b'y'^2 + 2g'x' + 2f'y' + c' = 0. \qquad (55.2)$$

First let us suppose that $a'b' \neq 0$. Then we can write

$$S' \equiv a'\left(x' + \frac{g'}{a'}\right)^2 + b'\left(y' + \frac{f'}{b'}\right)^2 + c' - \frac{g'^2}{a'} - \frac{f'^2}{b'} = 0,$$

and a translation to parallel axes through $(-g'/a', -f'/b')$ changes the equation of the curve into

$$\bar{S} \equiv \bar{a}\bar{x}^2 + \bar{b}\bar{y}^2 + \bar{c} = 0.$$

Since we have not multiplied across at any stage by a number, invariant theory yields

$$\bar{a} + \bar{b} = I; \quad \bar{a}\bar{b} = J; \quad \bar{a}\bar{b}\bar{c} = \Delta.$$

Hence $\bar{a}$, $\bar{b}$ are the roots λ_1, λ_2 of the quadratic equation

$$\lambda^2 - I\lambda + J = 0;$$

also $\bar{c} = \Delta/J$.

Dropping the bars from $\bar{x}$, $\bar{y}$ we may write the equation of the curve if $\Delta \neq 0$ in the form

$$\frac{x^2}{-\dfrac{\Delta}{\lambda_1 J}} + \frac{y^2}{-\dfrac{\Delta}{\lambda_2 J}} = 1. \qquad (55.3)$$

Consequently the curve is an ellipse (including a circle for $\lambda_1 = \lambda_2$) or a hyperbola. If $\Delta = 0$, it follows that $\bar{c} = 0$ and the curve is a line-pair with the equation

$$\lambda_1 x^2 + \lambda_2 y^2 = 0.$$

We must return to (55.2) and ask: "What happens to this equation if the rotation simultaneously reduces one of a' or b' to zero as well as making $h' = 0$." (The rotation cannot make $a' = b' = h' = 0$ for by invariant theory this implies $a + b = 0$ and $ab - h^2 = 0$. However, these equations cannot be satisfied for real values of a, b and h unless $a = b = h = 0$. This latter combination is excluded since S is quadratic.) There is no loss in generality if we select $a' = 0$. If $b' = 0$ we need only interchange the OX' and OY' axes. Equation (55.2) now becomes

$$S' \equiv b'y'^2 + 2g'x' + 2f'y' + c' = 0, \qquad (55.4)$$

and let us suppose that $g' \neq 0$. Then we can write

$$S' \equiv b'\left(y' + \frac{f'}{b'}\right)^2 + 2g'\left(x' - \frac{f'^2}{2b'g'} + \frac{c'}{2g'}\right) = 0$$

and a translation to parallel axes through $\left(\dfrac{f'^2}{2b'g'} - \dfrac{c'}{2g'}, \ -\dfrac{f'}{b'}\right)$ reduces the equation of the curve to

$$\bar{S} \equiv \bar{b}\bar{y}^2 + 2\bar{g}\bar{x} = 0.$$

By invariant theory we have $\bar{b} = I$ and $-\bar{b}\bar{g}^2 = \Delta$. Therefore, $\bar{g} = \pm\sqrt{-\Delta/I}$ and on dropping the bars the equation of the curve is

$$y^2 = 2\sqrt{-\frac{\Delta}{I^3}}\,x, \qquad (55.5)$$

which represents a parabola with semi-latus rectum $\sqrt{-\Delta/I^3}$. (Note the appearance of the *absolute* invariant Δ/I^3.)

Finally, we return to (55.4) and examine the case when $g' = 0$ by the rotation which causes $h' = a' = 0$. The equation is

$$S' \equiv b'y'^2 + 2f'y' + c' = 0,$$

which clearly factorizes and hence represents a line-pair.

Consequently we have proved that

$$ax^2 + 2hxy + by^2 + 2gx + 2fy + c = 0$$

represents either a line-pair, ellipse, hyperbola or parabola. We include all these curves under the term **conic.** In particular, we refer to the ellipse and hyperbola as **central conics,** whilst the line-pair is called a **degenerate conic.**

56. Central conic

In the last section we demonstrated that if

$$ax^2 + 2hxy + by^2 + 2gx + 2fy + c = 0$$

represents a central conic, new axes can be chosen so that its equation takes the standard form (55.3), namely

$$\frac{x^2}{-\dfrac{\Delta}{\lambda_1 J}} + \frac{y^2}{-\dfrac{\Delta}{\lambda_2 J}} = 1, \qquad (56.1)$$

where,

$$\lambda_1 + \lambda_2 = I; \quad \lambda_1\lambda_2 = J.$$

This curve represents an ellipse if $-\Delta/\lambda_1 J$ and $-\Delta/\lambda_2 J$ are both positive. In this case their product $\Delta^2/\lambda_1\lambda_2 J^2 = \Delta^2/J^3$ and the sum of their reciprocals $-J(\lambda_1 + \lambda_2)/\Delta = -JI/\Delta$ are both positive. Hence $J > 0$ and $I/\Delta < 0$.

Conversely, if $J > 0$ and $I/\Delta < 0$, then $IJ/\Delta < 0$. That is, $(\lambda_1 + \lambda_2)J/\Delta < 0$. But $(\lambda_1 J/\Delta)(\lambda_2 J/\Delta) = J^3/\Delta^2 > 0$, which means that $\lambda_1 J/\Delta$ and $\lambda_2 J/\Delta$ have the same sign. Hence they are both negative and consequently (56.1) represents an ellipse.

If $\Delta/\lambda_1 J$ and $\Delta/\lambda_2 J$ are both positive, the curve (56.1) has no real points and we call it a virtual ellipse. The necessary and sufficient conditions, by an argument almost identical with the preceding, are $J > 0$ and $I/\Delta > 0$.

The curve (56.1) represents a hyperbola if λ_1 and λ_2 have opposite signs and so $J = \lambda_1\lambda_2 < 0$. Conversely, if $J < 0$ then λ_1 and λ_2 have opposite signs and (56.1) must represent in this case a hyperbola.

57. Parabola

We have seen in §55 that $ax^2 + 2hxy + by^2 + 2gx + 2fy + c = 0$ when $\Delta \neq 0$ represents a parabola if a rotation exists which ensures $a' = h' = 0$ (or for that matter $b' = h' = 0$). That is, $a'b' - h'^2 = 0$ and so by invariant theory $J = 0$. It remains to show that this parabola is always real. To do this, verify the algebraic identities

$$a\Delta = (ac - g^2)J - (hg - af)^2,$$
$$b\Delta = (bc - f^2)J - (hf - bg)^2.$$

For the parabola $J = 0$ and addition yields

$$I\Delta = -[(hg - af)^2 + (hf - bg)^2] < 0.$$

Thus $\Delta/I^3 < 0$ and so we can extract a real square root in (55.5).

58. Line-pair

It has been proved that $ax^2 + 2hxy + by^2 + 2gx + 2fy + c = 0$ represents a line-pair when $\Delta = 0$. For the case $h' = 0$ but $a'b' \neq 0$, we saw in §55 that this line-pair reduces to $\lambda_1 x^2 + \lambda_2 y^2 = 0$. This equation represents real intersecting or complex conjugate intersecting lines according as $\lambda_1\lambda_2$ is less than or greater than zero respectively. That is, according as $J < 0$ or > 0 respectively.

If the rotation causes $h' = a' = 0$, and so $J = 0$, then from (55.4) we calculate that $\Delta' = -b'g'^2$ and so $g' = 0$ if $\Delta' = 0$. Consequently the line-pair has the equation $S' \equiv b'y'^2 + 2f'y' + c' = 0$, which represents two parallel lines. These lines are real and distinct, coincident or complex conjugate according as $b'c' - f'^2$ is less than, equal to, or greater than zero respectively. In this case there has been no translation, hence we can apply the invariant theory under the rotation to yield

$$a + b = b'; \quad c = c'; \quad g^2 + f^2 = f'^2.$$

Thus

$$b'c' - f'^2 = c(a + b) - (g^2 + f^2) = ca - g^2 + bc - f^2.$$

Hence the parallel line-pair is real and distinct, coincident or complex conjugate according as $ca - g^2 + bc - f^2$ is less than, equal to or greater than zero. (In the notation of §67, this expression is $\mathscr{A} + \mathscr{B}$.)

59. Classification of conic $ax^2 + 2hxy + by^2 + 2gx + 2fy + c = 0$

We now collect the results of sections 56, 57 and 58 in the following table:

Conic	Δ	J	Δ/I	$ca - g^2 + bc - f^2$
Real ellipse	$\neq 0$	+ve	−ve	—
Virtual ellipse	$\neq 0$	+ve	+ve	—
Hyperbola	$\neq 0$	−ve	—	—
Parabola	$\neq 0$	0	—	—
Real intersecting lines	0	−ve	—	—
Conjugate complex intersecting lines	0	+ve	—	—
Real distinct parallel lines	0	0	—	−ve
Conjugate complex parallel lines	0	0	—	+ve
Coincident lines	0	0	—	0

Miscellaneous examples

1. Obtain by invariant theory the standard form of the following conics:

(*a*) $x^2 - 4xy + y^2 + 8x + 2y - 5 = 0$,
(*b*) $3x^2 + 8xy - 3y^2 + 6x + 8y + 4 = 0$,
(*c*) $x^2 - 6xy + 9y^2 + 4x + 8y + 15 = 0$,
(*d*) $3x^2 - 24xy + 10y^2 + 6x + 52y = 0$,
(*e*) $3x^2 + 2xy + 3y^2 + 2x - 6y + 12\frac{1}{2} = 0$,
(*f*) $4x^2 - 24xy - 6y^2 + 4x - 12y + 1 = 0$.

2. Show that the equation of any conic through the four points $(\alpha, 0)$, $(\beta, 0)$, $(0, \gamma)$, $(0, \delta)$ can be written in the form

$$\frac{1}{\alpha\beta}(x - \alpha)(x - \beta) + \frac{1}{\gamma\delta}(y - \gamma)(y - \delta) - 1 + 2hxy = 0,$$

where h is a variable. In particular, find the eccentricity of the conic through the five points $(1, 0)$, $(2, 0)$, $(0, 1)$, $(0, 2)$ and $(2, 2)$.

3. Prove that all the conics $x^2 + y^2 + (x \cos\theta + y \sin\theta)^2 = 1$ have the same eccentricity $1/\sqrt{2}$.

4. A conic is given by the equation

$$(1 + \lambda^2)(x^2 + y^2) - 4\lambda xy + 2\lambda(x + y) + 2 = 0,$$

where λ may take any real value. Show that the conic is (*a*) a real ellipse with eccentricity $2\sqrt{\lambda}/(1 + \lambda)$ for $\lambda > \frac{1}{2}$ (but excluding $\lambda = 1$), (*b*) a parabola with the standard form $\sqrt{2}y^2 = x$ for $\lambda = 1$, (*c*) a real line-pair for $\lambda = \frac{1}{2}$ but a conjugate complex parallel line-pair for $\lambda = -1$.

5. If the conic $ax^2 + 2hxy + by^2 + 2gx + 2fy + c = 0$ is a rectangular hyperbola, show that its equation referred to its asymptotes as axes is $2xy + \Delta/(-J)^{3/2} = 0$.

6. If the conic $ax^2 + 2hxy + by^2 + 2gx + 2fy + c = 0$ is an ellipse, show that its area is $\pi\Delta/J^{3/2}$.

Answers

Miscellaneous examples. 1. (*a*) $\frac{x^2}{6} - \frac{y^2}{2} = 1$, (*b*) $\frac{x^2}{\frac{1}{5}} - \frac{y^2}{\frac{1}{5}} = -1$, (*c*) $y^2 = \frac{\sqrt{10}}{5}x$, (*d*) $\frac{x^2}{\frac{35}{6}} - \frac{y^2}{\frac{35}{19}} = 1$, (*e*) $\frac{x^2}{2} + \frac{y^2}{4} = -1$, (*f*) $7x^2 - 6y^2 = 0$. **2.** $\sqrt{2/5}$.

CHAPTER IX

LINE AT INFINITY

60. Homogeneous cartesian coordinates

For many purposes it is convenient to introduce three numbers X, Y, Z not all zero, defined by

$$X:Y:Z = x:y:1.$$

The numbers (X, Y, Z) are called **homogeneous cartesian coordinates** of the point (x, y). Since $\rho X:\rho Y:\rho Z = X:Y:Z$ we can also select $(\rho X, \rho Y, \rho Z)$ as the homogeneous coordinates of the point (x, y).

The equation of the straight line $lx + my + n = 0$ becomes $lX + mY + nZ = 0$ in homogeneous coordinates.

61. Point at infinity

The parametric equations of the straight line through A with gradient $\tan\psi$ (see §7, example 6) are $x = x_A + t\cos\psi$, $y = y_A + t\sin\psi$. The homogeneous coordinates of any point P on it are $P \equiv (x_A + t\cos\psi,\ y_A + t\sin\psi,\ 1)$ or equally well $P \equiv \left(\frac{x_A}{t} + \cos\psi,\ \frac{y_A}{t} + \sin\psi,\ \frac{1}{t}\right)$ obtained on division by t. The distance between P and A is t and as we allow t to tend to infinity the point P tends to be situated at an infinite distance from A. In the limit as t tends to infinity, the coordinates of P become $(\cos\psi, \sin\psi, 0)$ and we say that these are the coordinates of the **point at infinity** on the straight line. (The point at infinity is not to be regarded as having any physical existence. It is a mathematical concept of great power in the subsequent development of the geometry of the conic.) It is worthy of note that we arrive at the same point if we allow t to tend to negative infinity. Thus a straight line has only one point at infinity. From this point of view we can regard the straight line as a closed curve with one unaccessible point at infinity. The features to observe are (i) that the third coordinate of a point at infinity is zero, (ii) the ratio of the second coordinate to the first coordinate is the gradient of the straight line.

The coordinates of the point at infinity on the x-axis and y-axis are respectively $(1, 0, 0)$ and $(0, 1, 0)$. The triple $(0, 0, 0)$ has been excluded by the definition of homogeneous coordinates.

62. Straight lines

The two straight lines $l_1x + m_1y + n_1 = 0$ and $l_2x + m_2y + n_2 = 0$ intersect at the unique point $\left(\dfrac{m_1n_2 - m_2n_1}{l_1m_2 - l_2m_1}, \dfrac{n_1l_2 - n_2l_1}{l_1m_2 - l_2m_1}\right)$ provided that $l_1m_2 - l_2m_1 \neq 0$. Using homogeneous coordinates, the two straight lines $l_1X + m_1Y + n_1Z = 0$ and $l_2X + m_2Y + n_2Z = 0$ intersect at the point $(m_1n_2 - m_2n_1, n_1l_2 - n_2l_1, l_1m_2 - l_2m_1)$. If the straight lines are parallel $l_1m_2 - l_2m_1 = 0$, from which we have $m_1n_2 - m_2n_1 = m_1n_2 - \dfrac{l_2m_1n_1}{l_1} = \dfrac{m_1}{l_1}(n_2l_1 - n_1l_2)$ and so the point of intersection is $(m_1, -l_1, 0)$ or for that matter also $(m_2, -l_2, 0)$. This point is the point at infinity on both the straight lines $l_1X + m_1Y + n_1Z = 0$ and $l_2X + m_2Y + n_2Z = 0$. Hence two parallel straight lines intersect in their common point at infinity. The strength of the homogeneous coordinate interpretation lies in the fact that there is now no exception to the statement that two straight lines intersect in one point. If the lines are parallel, this point of intersection is their common point at infinity.

Now consider three straight lines given in homogeneous coordinates by the equations $l_1X + m_1Y + n_1Z = 0$, $l_2X + m_2Y + n_2Z = 0$ and $l_3X + m_3Y + n_3Z = 0$. In §10, the necessary condition for the concurrency of the three lines was not a sufficient condition. If the condition is satisfied the lines are either concurrent or parallel to one another. With the concept "point at infinity" we have that equation (10.4) is now the necessary and sufficient condition that the three given straight lines be concurrent. The case of parallel lines is automatically included as intersecting at their common point at infinity.

In §5 we saw that the points $P \equiv (\lambda x_B + \mu x_A, \lambda y_B + \mu y_A, \lambda + \mu)$ and $Q = (\lambda x_B - \mu x_A, \lambda y_B - \mu y_A, \lambda - \mu)$ are harmonic conjugates with respect to A and B. Put $\lambda = \mu$, then $P = (x_B + x_A, y_B + y_A, 2)$ and $Q = (x_B - x_A, y_B - y_A, 0)$. P is the mid-point of AB and Q is the point at infinity on AB. Thus the point at infinity is the harmonic conjugate of the mid-point of AB with respect to A and B.

The third coordinate Z of a point at infinity is always zero. In order to preserve the theorem that all linear equations represent straight lines, we agree to say that the equation $Z = 0$ represents the straight line at infinity. The reader must not picture an actual straight line at infinity. We have merely agreed to give the name "line at infinity" to the collection of all points whose third homogeneous coordinate is zero. No meaning is attached to the angle between two straight lines one of which is the line at infinity.

63. Circular points at infinity

Consider the circle whose equation is

$$ax^2 + ay^2 + 2gx + 2fy + c = 0.$$

Introducing homogeneous coordinates this becomes

$$aX^2 + aY^2 + 2gXZ + 2fYZ + cZ^2 = 0. \qquad (63.1)$$

We now ask: "Has the circle any points at infinity?" That is, are there any points on the circle whose third homogeneous coordinate is zero? The answer is provided by the solution of the two equations

$$aX^2 + aY^2 + 2gXZ + 2fYZ + cZ^2 = 0; \quad Z = 0,$$

which simplify to $X^2 + Y^2 = 0$ and $Z = 0$. These equations have two homogeneous solutions $(1, i, 0)$ and $(1, -i, 0)$. These points are called the **circular points at infinity**,† and are generally denoted by I and J. The terms $I = a + b$ and $J = ab - h^2$ have already been introduced in the theory of invariants. No confusion with the circular points I and J will arise. The meaning of I and J will always be clear from the text. The remarkable fact is that these points are the same for all circles. That is, the straight line at infinity intersects all circles in the same two points at infinity.

Conversely, we shall now prove that any conic with *real coefficients*, passing through *one* of the circular points at infinity is a circle. The general equation of a conic in homogeneous coordinates is

$$aX^2 + 2hXY + bY^2 + 2gXZ + 2fYZ + cZ^2 = 0.$$

The conic passes through $(1, i, 0)$ and so $a + 2hi - b = 0$. Hence $a = b$, $h = 0$ and the conic is a circle. If a conic passes through *both* circular points at infinity, $a + 2hi - b = 0$ and $a - 2hi - b = 0$. In this case $a = b$ and $h = 0$ irrespective of the reality of the coefficients.

By retaining a in equation (63.1) we may now ask for the interpretation when $a = 0$. The circle degenerates to $Z(2gX + 2fY + cZ) = 0$. That is, a straight line together with the straight line at infinity must be regarded as a special case of a circle. Of course, the centre at $(-g/a, -f/a)$ and radius $\sqrt{g^2 + f^2 - ac}/a$ all tend to infinity.

Example 1. Show that the line-pair joining the origin to the circular points is $x^2 + y^2 = 0$.

---

† We invariably reserve the letter i to denote the solution of the equation $i^2 + 1 = 0$. We cannot distinguish between $+\sqrt{-1}$ and $-\sqrt{-1}$. If i takes one of these values, then $-i$ is the other.

64. Conic and line at infinity

In homogeneous coordinates the equation (29.1) of the ellipse becomes

$$\frac{X^2}{a^2} + \frac{Y^2}{b^2} - Z^2 = 0$$

and so the intersections with the line at infinity are given by

$$\frac{X^2}{a^2} + \frac{Y^2}{b^2} - Z^2 = 0; \quad Z = 0.$$

Thus the line at infinity intersects the ellipse in the two *complex conjugate* points $(a, bi, 0)$ and $(a, -bi, 0)$.

The homogeneous equation of the hyperbola (37.1) is

$$\frac{X^2}{a^2} - \frac{Y^2}{b^2} - Z^2 = 0.$$

Hence the line at infinity intersects the hyperbola in the two *real* points $(a, b, 0)$ and $(a, -b, 0)$.

Further, the homogeneous equation of the parabola (44.1) is

$$Y^2 - 4aXZ = 0$$

and the line at infinity intersects the parabola at the *coincident* points $(1, 0, 0)$. That is, the line at infinity touches the parabola at the point at infinity on its axis.

Consequently we have demonstrated that a conic is an ellipse, parabola or hyperbola according as its intersections with the line at infinity are complex conjugate, coincident or real and distinct respectively.

Example 2. By considering the points of intersection of the conic $aX^2 + 2hXY + bY^2 + 2gXZ + 2fYZ + cZ^2 = 0$, $(\Delta \neq 0)$, and the line at infinity show that the conic is an ellipse, parabola or hyperbola according as $J > 0$, $= 0$ or < 0 respectively.

65. Joachimsthal's section formula

The coordinates of the point P which divides AB in the ratio λ/μ are $((\mu x_A + \lambda x_B)/(\mu + \lambda), (\mu y_A + \lambda y_B)/(\mu + \lambda))$. If we use homogeneous coordinates $P \equiv \left(\frac{\mu X_A}{Z_A} + \frac{\lambda X_B}{Z_B}, \frac{\mu Y_A}{Z_A} + \frac{\lambda Y_B}{Z_B}, \mu + \lambda\right)$.
That is, $P \equiv (\mu X_A Z_B + \lambda X_B Z_A, \mu Y_A Z_B + \lambda Y_B Z_A, (\mu + \lambda)Z_A Z_B)$.
We substitute $\lambda Z_A = \nu$, $\mu Z_B = \rho$; then

$$P \equiv (\rho X_A + \nu X_B, \ \rho Y_A + \nu Y_B, \ \rho Z_A + \nu Z_B). \quad (65.1)$$

The ratio $\nu/\rho = (Z_A/Z_B)(\lambda/\mu)$ and so $\nu/\rho = \lambda/\mu = AP/PB$ if and only if $Z_A = Z_B$. If A and B in Joachimsthal's section formula are finite points, it is sometimes convenient to choose $Z_A = Z_B = 1$. Then $\nu/\rho = AP/PB$. If either A or B is a point at infinity, the ratio λ/μ has no geometrical significance; yet the point (65.1) traves out all points of the straight line AB as ν/ρ ranges from $-\infty$ to $+\infty$.

CHAPTER X

CONIC

66. Conic

In this chapter we discuss the properties of the conic defined by $ax^2 + 2hxy + by^2 + 2gx + 2fy + c = 0$. This equation has five effective constants, namely the ratios of the six quantities a, b, c, , g and h. The condition that a point lie on a conic is linear in these coefficients. In general, one unique conic passes through five given points, because five simultaneous linear equations have to be satisfied by the five ratios $a{:}b{:}c{:}f{:}g{:}h$. If three of the five points are collinear, the conic is degenerate but unique, consisting of the line joining the three collinear points and the line joining the other two points. However, if four of the points are collinear on a line l, the conic is no longer determined uniquely but is degenerate, consisting of the line l and any line through the fifth point.

Example 1. Obtain the equation of the conic through the five points: (0, 0), (0, 1), (1, 0), (1, 1) and (− 1, 3).

Example 2. How many parabolae pass through four points? Find the equations of the parabolae through the four points: (0, 0), (8, 0), (0, 8) and (− 1, 3).

67. Cofactor notation

The reader familiar with determinants need only note that in this section we introduce the cofactors in the determinant Δ, and list their properties. The reader who has not had an elementary course in determinants should note the following: the six quantities $\mathscr{A}$, $\mathscr{B}$, $\mathscr{C}$, $\mathscr{F}$, $\mathscr{G}$ and $\mathscr{H}$ are defined by

$$\left.\begin{array}{lll} \mathscr{A} = bc - f^2; & \mathscr{B} = ca - g^2; & \mathscr{C} = ab - h^2; \\ \mathscr{F} = gh - af; & \mathscr{G} = hf - bg; & \mathscr{H} = fg - ch. \end{array}\right\} \qquad (67.1)$$

The reader is asked to verify the following two sets of algebraic identities:

$$\left.\begin{array}{l} a\mathscr{A} + h\mathscr{H} + g\mathscr{G} = \Delta;\ h\mathscr{A} + b\mathscr{H} + f\mathscr{G} = 0;\ g\mathscr{A} + f\mathscr{H} + c\mathscr{G} = 0; \\ a\mathscr{H} + h\mathscr{B} + g\mathscr{F} = 0;\ h\mathscr{H} + b\mathscr{B} + f\mathscr{F} = \Delta;\ g\mathscr{H} + f\mathscr{B} + c\mathscr{F} = 0; \\ a\mathscr{G} + h\mathscr{F} + g\mathscr{C} = 0;\ h\mathscr{G} + b\mathscr{F} + f\mathscr{C} = 0;\ g\mathscr{G} + f\mathscr{F} + c\mathscr{C} = \Delta. \end{array}\right\} \qquad (67.2)$$

$$\left.\begin{aligned}\mathscr{BC}-\mathscr{F}^2&=a\Delta;\ \mathscr{CA}-\mathscr{G}^2=b\Delta;\ \mathscr{AB}-\mathscr{H}^2=c\Delta;\\ \mathscr{GH}-\mathscr{AF}&=f\Delta;\ \mathscr{HF}-\mathscr{BG}=g\Delta;\ \mathscr{FG}-\mathscr{CH}=h\Delta.\end{aligned}\right\}\quad(67.3)$$

Example 3. Establish the following results:

(a) $a\mathscr{G}^2 + 2h\mathscr{F}\mathscr{G} + b\mathscr{F}^2 + 2g\mathscr{C}\mathscr{G} + 2f\mathscr{C}\mathscr{F} + c\mathscr{C}^2 = \mathscr{C}\Delta$,

(b) $\begin{vmatrix} \mathscr{A} & \mathscr{H} & \mathscr{G} \\ \mathscr{H} & \mathscr{B} & \mathscr{F} \\ \mathscr{G} & \mathscr{F} & \mathscr{C} \end{vmatrix} = \Delta^2$,

(c) $b(ax + hy + g)^2 - 2h(ax + hy + g)(hx + by + f) + a(hx + by + f)^2$
$= \mathscr{C}(ax^2 + 2hxy + by^2 + 2gx + 2fy + c) - \Delta.$

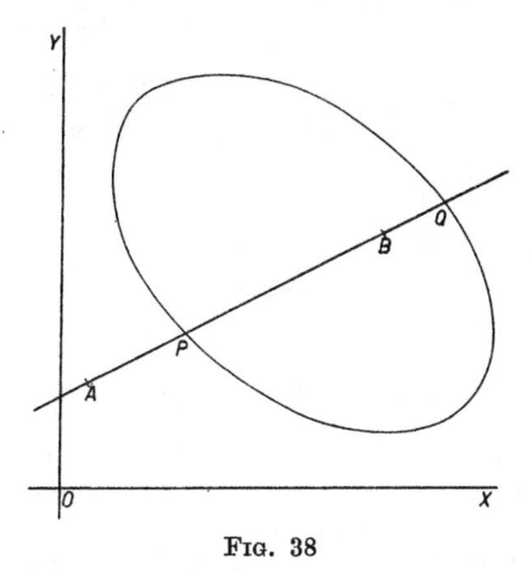

FIG. 38

68. Joachimsthal's quadratic equation

We now investigate the points of intersection of a straight line AB and the conic

$$ax^2 + 2hxy + by^2 + 2gx + 2fy + c = 0,$$

where the coefficients a, b, c, f, g and h are all real. Using homogeneous coordinates, this equation becomes

$$aX^2 + 2hXY + bY^2 + 2gXZ + 2fYZ + cZ^2 = 0.$$

It will be convenient now to use x, y, z instead of X, Y, Z as homogeneous coordinates. If we require at any stage to revert to non-homogeneous coordinates, it is only necessary to put $z = 1$. Let us then write the equation of the conic

$$S \equiv S(x, y, z) \equiv ax^2 + 2hxy + by^2 + 2gzx + 2fyz + cz^2 = 0. \tag{68.1}$$

By (65.1) the coordinates of P (Fig. 38) which divides AB in the ratio $\lambda:\mu$ are (if $z_A = z_B$)

$$P \equiv (\mu x_A + \lambda x_B, \mu y_A + \lambda y_B, \mu z_A + \lambda z_B).$$

This point P lies on the conic if

$$a(\mu x_A + \lambda x_B)^2 + 2h(\mu x_A + \lambda x_B)(\mu y_A + \lambda y_B) + b(\mu y_A + \lambda y_B)^2 + 2g(\mu z_A + \lambda z_B)(\mu x_A + \lambda x_B) + 2f(\mu y_A + \lambda y_B)(\mu z_A + \lambda z_B) + c(\mu z_A + \lambda z_B)^2 = 0.$$

That is

$$\mu^2 S_A + 2\mu\lambda T_{AB} + \lambda^2 S_B = 0, \tag{68.2}$$

where†

$$\left.\begin{aligned} S_A &\equiv S(x_A, y_A, z_A); \quad S_B \equiv S(x_B, y_B, z_B); \\ T_{AB} \equiv T_{BA} &\equiv ax_Ax_B + h(x_Ay_B + x_By_A) + by_Ay_B + g(z_Ax_B \\ &\quad + x_Az_B) + f(y_Az_B + y_Bz_A) + cz_Az_B, \\ &\equiv x_A(ax_B + hy_B + gz_B) + y_A(hx_B + by_B + fz_B) \\ &\quad + z_A(gx_B + fy_B + cz_B), \\ &\equiv x_B(ax_A + hy_A + gz_A) + y_B(hx_A + by_A + fz_A) \\ &\quad + z_B(gx_A + fy_A + cz_A). \end{aligned}\right\} \tag{68.3}$$

Equation (68.2), known as Joachimsthal's quadratic equation, has two roots in the ratio λ/μ, and they correspond to two points of intersection of the straight line AB and the conic.

We have established the fundamental theorem that a straight line and a conic intersect in two points. These points may be real and distinct as in the figure or coincident if AB touches the conic. If AB does not cut the conic in real points, equation (67.2) has complex conjugate points.

An exception occurs if the conic degenerates to the line-pair consisting of AB and any other straight line. Then the number of intersections of AB and the degenerate conic are infinite and so equation (68.2) is an identity. Since $S_A = S_B = 0$, we have

$$T_{AB} \equiv x_B(ax_A + hy_A + gz_A) + y_B(hx_A + by_A + fz_A) + z_B(gx_A + fy_A + cz_A) \equiv 0$$

for an infinite number of positions of the point B. Consequently

$$\begin{aligned} ax_A + hy_A + gz_A &= 0, \\ hx_A + by_A + fz_A &= 0, \\ gx_A + fy_A + cz_A &= 0. \end{aligned}$$

† In more compact form, we may write

$$T_{AB} = \frac{1}{2}\left[x_A\frac{\partial S_B}{\partial x_B} + y_A\frac{\partial S_B}{\partial y_B} + z_A\frac{\partial S_B}{\partial z_B}\right] = \frac{1}{2}\left[x_B\frac{\partial S_A}{\partial x_A} + y_B\frac{\partial S_A}{\partial y_A} + z_B\frac{\partial S_A}{\partial z_A}\right].$$

The elimination of the ratios $x_A : y_A : z_A$ yields the result (previously proved in section 15) that the conic $S = 0$ is degenerate if

$$\Delta \equiv \begin{vmatrix} a & h & g \\ h & b & f \\ g & f & c \end{vmatrix} = 0.$$

Example 4. Obtain the coordinates of the points of intersection of the conic $2x^2 + xy + 2y^2 - 6x - 6y + 4 = 0$ and the line joining (*a*) (3, 4) and (− 1, − 4), (*b*) (3, 1) and (1, 3), (*c*) (0, − 1) and (2, 1).

69. Polar properties

Two points A and B are said to be **conjugate** with respect to the conic $S = 0$ if the two points of intersection (Fig. 38) P and Q divide A and B harmonically. Then $AP/PB = -AQ/QB$ and so the roots of (68.2) are equal in magnitude but opposite in sign. That is, the sum of the roots is zero. Thus the necessary and sufficient condition that A and B be conjugate points with respect to the conic $S = 0$ is $T_{AB} = 0$.

The **polar** of a point A with respect to the conic $S = 0$ is defined to be the locus of all points B which are conjugate to A. Hence the polar of A is given by $T_A = 0$, where

$$T_A \equiv (ax_A + hy_A + gz_A)x + (hx_A + by_A + fz_A)y + (gx_A + fy_A + cz_A)z \tag{69.1}$$

or in the alternative† form

$$T_A \equiv (ax + hy + gz)x_A + (hx + by + fz)y_A + (gx + fy + cz)z_A. \tag{69·2}$$

This equation is linear in x, y and z and so represents a straight line.

Conversely, the **pole** of a straight line is defined to be that point whose polar is the given straight line. Let us obtain the coordinates of the pole A of the straight line $lx + my + nz = 0$. The polar of A is given by (69.1) which must then represent the same straight line as $lx + my + nz = 0$. The comparison of coefficients yields

$$ax_A + hy_A + gz_A = \rho l, \tag{69.3}$$

$$hx_A + by_A + fz_A = \rho m, \tag{69.4}$$

$$gx_A + fy_A + cz_A = \rho n, \tag{69.5}$$

† We may also write

$$T_A \equiv \frac{1}{2}\left[x_A \frac{\partial S}{\partial x} + y_A \frac{\partial S}{\partial y} + z_A \frac{\partial S}{\partial z}\right] = \frac{1}{2}\left[x \frac{\partial S_A}{\partial x_A} + y \frac{\partial S_A}{\partial y_A} + z \frac{\partial S_A}{\partial z_A}\right].$$

where ρ is some factor of proportionality. Multiply (69.3) by $\mathscr{A}$, (69.4) by $\mathscr{H}$, (69.5) by $\mathscr{G}$ and add the three equations. In virtue of (67.2) we obtain

$$\Delta x_A = \rho(l\mathscr{A} + m\mathscr{H} + n\mathscr{G}).$$

Next repeat this operation with the multipliers $\mathscr{H}$, $\mathscr{B}$, $\mathscr{F}$ and finally repeat again with the multipliers $\mathscr{G}$, $\mathscr{F}$, $\mathscr{C}$ and the results are

$$\Delta y_A = \rho(l\mathscr{H} + m\mathscr{B} + n\mathscr{F}),$$
$$\Delta z_A = \rho(l\mathscr{G} + m\mathscr{F} + n\mathscr{C}).$$

Hence the pole of the straight line $lx + my + nz = 0$ is at the point

$$(l\mathscr{A} + m\mathscr{H} + n\mathscr{G}, l\mathscr{H} + m\mathscr{B} + n\mathscr{F}, l\mathscr{G} + m\mathscr{F} + n\mathscr{C}). \quad (69.6)$$

Consider the two straight lines $l_1x + m_1y + n_1z = 0$ and $l_2x + m_2y + n_2z = 0$. The pole of $l_1x + m_1y + n_1z = 0$ is at the point $(l_1\mathscr{A} + m_1\mathscr{H} + n_1\mathscr{G}, l_1\mathscr{H} + m_1\mathscr{B} + n_1\mathscr{F}, l_1\mathscr{G} + m_1\mathscr{F} + n_1\mathscr{C})$. This point lies on $l_2x + m_2y + n_2z = 0$ if

$$l_1l_2\mathscr{A} + m_1m_2\mathscr{B} + n_1n_2\mathscr{C} + (m_1n_2 + m_2n_1)\mathscr{F}$$
$$+ (n_1l_2 + n_2l_1)\mathscr{G} + (l_1m_2 + l_2m_1)\mathscr{H} = 0. \quad (69.7)$$

This relation is symmetrical in the indices 1 and 2. Thus the pole of $l_2x + m_2y + n_2z = 0$ lies on $l_1x + m_1y + n_1z = 0$. When two lines are so situated that each passes through the pole of the other, they are called **conjugate lines** with respect to the conic. The necessary and sufficient condition for the conjugacy of two straight lines is that equation (69.7) be satisfied.

Example 5. Find the equations of the polars of the points (3, 0), (0, $-$ 2) and (6, 2) with respect to the conic $16x^2 + 4xy + 19y^2 - 56x - 72y + 84 = 0$. Further prove that these three straight lines are concurrent.

Example 6. Obtain the coordinates of the pole of the straight line $42x + 19y = 76$ with respect to the conic of example 5. Further, is this straight line conjugate to the straight lines (a) $x + y = 1$, (b) $x - y = 2$, (c) $x + y = 0$, (d) $2x + 3y = 1$.

70. Tangent properties

We return to Joachimsthal's equation (68.2), namely

$$\mu^2S_A + 2\mu\lambda T_{AB} + \lambda^2S_B = 0. \quad (70.1)$$

Suppose A lies on the conic; then $S_A = 0$.

We define the tangent at A to be the limiting position of the secant AL through L (Fig. 39) as L tends to coincidence with A. Consequently the tangent at A cuts the conic $S = 0$ in coincident points, and so Joachimsthal's equation $2\mu\lambda T_{AB} + \lambda^2 S_B = 0$ has coincident roots $\lambda/\mu = 0$. Therefore, $T_{AB} = 0$ if B lies on the tangent at A. That is, the tangent at A has the equation

$$T_A \equiv (ax_A + hy_A + gz_A)x + (hx_A + by_A + fz_A)y + (gx_A + fy_A + cz_A)z = 0. \quad (70.2)$$

It is identical with (69.1) which gives the polar of A. Hence the pole of a tangent line is at its point of contact.

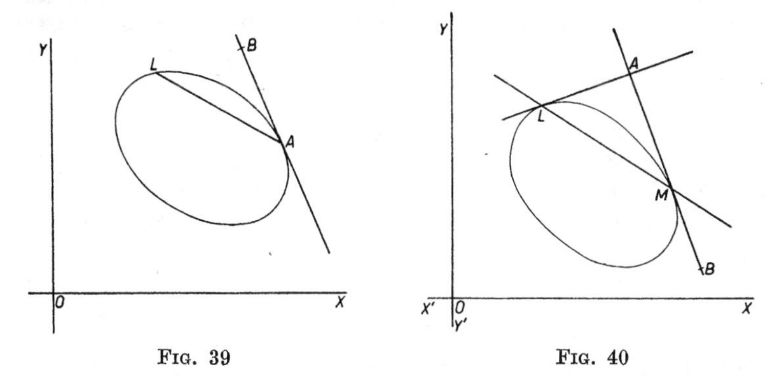

FIG. 39 FIG. 40

The straight line $lx + my + nz = 0$ is a tangent to the conic $S = 0$ if its pole, given by the coordinates (69.6) lies on it. Hence the necessary and sufficient condition that $lx + my + nz = 0$ be tangent to the conic $S = 0$ is

$$l^2\mathscr{A} + m^2\mathscr{B} + n^2\mathscr{C} + 2mn\mathscr{F} + 2nl\mathscr{G} + 2lm\mathscr{H} = 0. \quad (70.3)$$

We can also say that a tangent line $lx + my + nz = 0$ is self-conjugate and this also yields the condition (70.3).

However, if AB is a tangent, but neither A nor B lies on the conic (Fig. 40), then the roots of (70.1) must coincide. That is, $S_A S_B - T_{AB}^2 = 0$. Thus the locus of the point B if A is held fixed is given by

$$S_A S - T_A^2 = 0. \quad (70.4)$$

This is a quadratic equation and must therefore represent the pair of tangents from A to the conic.

The polar of A, given by $T_A = 0$, cuts the conic $S = 0$ where $T_A = S = 0$. These two points also lie on the line-pair (70.4). Accordingly the polar of A is the line LM (Fig. 40) joining the points of contact of the tangents from A to the conic.

Example 7. Eliminate z between the equations $lx + my + nz = 0$ and $ax^2 + 2hxy + by^2 + 2gxz + 2fyz + cz^2 = 0$. Express the condition that the resulting quadratic in x/y has equal roots. Hence show that (70.3) is the tangency condition.

Example 8. Verify that the equation $S_AS - T_A{}^2 = 0$ represents two straight lines.

Example 9. Calculate the angle between the pair of tangents from the point (1, 2) to the conic $3x^2 + 8xy - 3y^2 + 6x + 8y + 4 = 0$.

71. Orthoptic locus

Let us find the locus of the points of intersection of perpendicular tangents to the conic $S = 0$. The pair of tangents from A to the conic is given by $S_AS - T_A{}^2 = 0$. Using non-homogeneous coordinates, we can write this

$$S_A(ax^2 + 2hxy + by^2 + 2gx + 2fy + c)$$
$$-\{(ax_A + hy_A + g)x + (hx_A + by_A + f)y + (gx_A + fy_A + c)\}^2 = 0.$$

This line-pair is mutually orthogonal if the sum of the coefficients of x^2 and y^2 is zero. Hence

$$(a + b)(ax_A{}^2 + 2hx_Ay_A + by_A{}^2 + 2gx_A + 2fy_A + c)$$
$$- (ax_A + hy_A + g)^2 - (hx_A + by_A + f)^2 = 0,$$

which can be written

$$\mathscr{C}(x_A{}^2 + y_A{}^2) - 2\mathscr{G}x_A - 2\mathscr{F}y_A + \mathscr{A} + \mathscr{B} = 0.$$

Consequently the locus of A is the circle

$$\Theta \equiv \mathscr{C}(x^2 + y^2) - 2\mathscr{G}x - 2\mathscr{F}y + \mathscr{A} + \mathscr{B} = 0. \qquad (71.1)$$

called the **orthoptic circle.**

The conic $S = 0$ is a parabola if $\mathscr{C} \equiv J = 0$ and then the circle (71.1) becomes the straight line

$$2\mathscr{G}x + 2\mathscr{F}y = \mathscr{A} + \mathscr{B}. \qquad (71.2)$$

We saw in §45 that this locus is the directrix of the parabola. In practice, this method is convenient if we wish to find the directrix of a given parabola.

Example 10. Show that the centre of the orthoptic circle is at $(\mathscr{G}, \mathscr{F}, \mathscr{C})$ and that its radius is $\sqrt{-I\Delta/J^2}$.

Example 11. Obtain the equation of the directrix of the parabola $x^2 - 6xy + 9y^2 - 2x - 3y + 1 = 0$.

72. Centre

We define the centre as the pole of the line at infinity $z = 0$ with respect to the conic $S = 0$. For the line at infinity $l = m = 0$, $n = 1$ and so by (69.6) its pole is at $(\mathscr{G}, \mathscr{F}, \mathscr{C})$ and coincides with the centre of the orthoptic circle.

If the conic is a parabola $\mathscr{C} \equiv J = 0$ and so the centre is the point at infinity $(\mathscr{G}, \mathscr{F}, 0)$.

Let any line through the centre C cut the conic at P and Q and the line at infinity in R. The line at infinity is the polar of C; hence the points C and R are conjugate to P and Q. Consequently C and R divide P and Q harmonically. It follows that (see §62) C is the mid-point of PQ. Thus all chords through the centre are bisected there.

From (67.2) we have $a\mathscr{G} + h\mathscr{F} + g\mathscr{C} = 0$ and $h\mathscr{G} + b\mathscr{F} + f\mathscr{C} = 0$. Therefore the centre lies on both the straight lines

$$ax + hy + gz = 0, \tag{72.1}$$

$$hx + by + fz = 0. \tag{72.2}$$

These straight lines are parallel if the conic is a parabola.

Example 12. Find the coordinates of the centre of the following conics:

(a) $3x^2 + 8xy - 3y^2 + 6x + 8y + 4 = 0$,

(b) $x^2 - 4xy + y^2 + 8x + 2y - 5 = 0$,

(c) $x^2 - 2xy + 2y^2 - 3x + 7y - 1 = 0$.

73. Conjugate diameters

Any line through the centre is called a **diameter.** Since the centre lies on both the straight lines (72.1) and (72.2) the equation of any diameter can be written

$$ax + hy + gz + \lambda(hx + by + fz) = 0. \tag{73.1}$$

By (69.6) the pole of this diameter is at the point

$$\begin{aligned}([a + \lambda h]\mathscr{A} + [h + \lambda b]\mathscr{H} + [g + \lambda f]\mathscr{G},\\ [a + \lambda h]\mathscr{H} + [h + \lambda b]\mathscr{B} + [g + \lambda f]\mathscr{F},\\ [a + \lambda h]\mathscr{G} + [h + \lambda b]\mathscr{F} + [g + \lambda f]\mathscr{C}).\end{aligned}$$

In virtue of the relations (67.2) this point has the coordinates $(\Delta, \lambda\Delta, 0)$. That is, the pole of the diameter (73.1) is at the point at infinity $(1, \lambda, 0)$.

Excluding the case of the parabola, we define **conjugate diameters** to be conjugate lines with respect to the conic.

Consider the two diameters

$$ax + hy + gz + \lambda_1(hx + by + fz) = 0,$$
$$ax + hy + gz + \lambda_2(hx + by + fz) = 0.$$

The pole of the first diameter is at $(1, \lambda_1, 0)$ and this point lies on the second diameter if

$$a + h(\lambda_1 + \lambda_2) + b\lambda_1\lambda_2 = 0.$$

The gradients γ_1 and γ_2 of these diameters are

$$\gamma_1 = -\frac{a + h\lambda_1}{h + b\lambda_1}; \quad \gamma_2 = -\frac{a + h\lambda_2}{h + b\lambda_2}.$$

Thus $\gamma_1 = \lambda_2$ and $\gamma_2 = \lambda_1$. Hence the gradients γ_1 and γ_2 of conjugate diameters satisfy the condition

$$a + h(\gamma_1 + \gamma_2) + b\gamma_1\gamma_2 = 0. \tag{73.2}$$

The mid-points of a system of parallel chords of a central conic are conjugate to their common point at infinity. Consequently the locus of the mid-points is the polar of this point at infinity. This polar line is the diameter conjugate to the diameter parallel to the chords. Thus the locus of the mid-points of a system of chords parallel to a diameter is the conjugate diameter.

Consider a point A on a diameter and the point at infinity which we shall for the moment denote by B on the conjugate diameter. The polar of B is the diameter through A. Thus A and B are conjugate points. Hence the polar of A passes through B. That is, the polar of A is parallel to the conjugate diameter.

Example 13. Show that $x - 3y + 2 = 0$ is a diameter of the conic $3x^2 - 2xy - y^2 + 2x + y - 1 = 0$. Obtain the conjugate diameter.

74. Principal axes

Principal axes are defined for the central conic to be a pair of mutually orthogonal conjugate diameters. The gradients γ_1 and γ_2 of the principal axes then satisfy the two equations

$$\left.\begin{aligned} a + h(\gamma_1 + \gamma_2) + b\gamma_1\gamma_2 &= 0, \\ \gamma_1\gamma_2 &= -1. \end{aligned}\right\} \tag{74.1}$$

Accordingly $\gamma_1 + \gamma_2 = -(a - b)/h$ and so γ_1 and γ_2 are the roots of the quadratic equation

$$\gamma^2 + \frac{a - b}{h}\gamma - 1 = 0.$$

The rotation which removes the product term in xy is given in §55 by $\tan 2\theta = 2h/(a - b)$. Hence the quadratic equation becomes

$$\gamma^2 + 2\gamma \cot 2\theta - 1 = 0,$$

and its roots are $\gamma_1 = \tan \theta$, $\gamma_2 = -\cot \theta$.

Thus, in general, there is only one pair of principal axes. Since the gradient of one of these axes is $\tan \theta$, we see that they lie along the major and minor axes in the case of an ellipse and along the transverse and conjugate axes if the conic is a hyperbola.

For the special case of the circle, $a = b$ and $h = 0$. Then equations (74.1) reduce to the one equation $\gamma_1\gamma_2 = -1$. That is, every pair of mutually orthogonal diameters of a circle are principal axes.

The following is a useful method of finding the equation of the principal axes. We proved in the last section that the polar of a point on a diameter is parallel to the conjugate diameter. Thus the polar of a point on a principal axis of a conic is perpendicular to that axis.

Let A be a point on a principal axis, then the axis is the straight line joining A to the centre $(\mathscr{G}, \mathscr{F}, \mathscr{C})$ of the conic and so its equation can be verified to be

$$(\mathscr{C}y_A - \mathscr{F})(x - x_A) - (\mathscr{C}x_A - \mathscr{G})(y - y_A) = 0.$$

This line is perpendicular to the polar of A given by (69.1). Hence

$$(\mathscr{C}y_A - \mathscr{F})(ax_A + hy_A + g) - (\mathscr{C}x_A - \mathscr{G})(hx_A + by_A + f) = 0.$$

Consequently the principal axes are

$$(\mathscr{C}y - \mathscr{F})(ax + hy + g) - (\mathscr{C}x - \mathscr{G})(hx + by + f) = 0. \quad (74.2)$$

The reader is asked to verify, using the standard form $y^2 = 4ax$ that the polar of a point on the axis of a parabola is also perpendicular to the axis. So we put $\mathscr{C} \equiv J = 0$ in (74.2) and we obtain that

$$\mathscr{F}(ax + hy + g) - \mathscr{G}(hx + by + f) = 0 \quad (74.3)$$

is the equation of the axis if $S = 0$ is a parabola.

The vertex of a parabola can be obtained as the point of intersection of the axis and the tangent at the vertex. This tangent is parallel to the directrix (71.2) and so its equation is $\mathscr{G}x + \mathscr{F}y + \lambda = 0$. But this line is a tangent and so by (70.3) we have

$$\mathscr{A}\mathscr{G}^2 + \mathscr{B}\mathscr{F}^2 + 2\mathscr{F}\mathscr{G}\mathscr{H} + 2\lambda(\mathscr{F}^2 + \mathscr{G}^2) = 0,$$

whence λ can be calculated, etc.

Example 14. Obtain the principal axes of the conics in Example 12.

75. Asymptotes

The **asymptotes** of a non-degenerate conic are defined as the pair of tangent lines from the centre to the conic. The centre of the conic is at $(\mathscr{G}, \mathscr{F}, \mathscr{C})$ and so the pair of tangents to the conic is obtained by substitution in (70.4). The result is

$$(a\mathscr{G}^2 + 2h\mathscr{F}\mathscr{G} + b\mathscr{F}^2 + 2g\mathscr{G}\mathscr{C} + 2f\mathscr{F}\mathscr{C} + c\mathscr{C}^2)S$$
$$- \{[a\mathscr{G} + h\mathscr{F} + g\mathscr{C}]x + [h\mathscr{G} + b\mathscr{F} + f\mathscr{C}]y + [g\mathscr{C} + f\mathscr{F} + c\mathscr{C}]z\}^2 = 0.$$

That is, in virtue of (67.2)

$$(\mathscr{G}[a\mathscr{G} + h\mathscr{F} + g\mathscr{C}] + \mathscr{F}[h\mathscr{G} + b\mathscr{F} + f\mathscr{C}] + \mathscr{C}[g\mathscr{G} + f\mathscr{F} + c\mathscr{C}])S - \Delta^2 z^2 = 0,$$

which reduces further to $\mathscr{C}\Delta S - \Delta^2 z^2 = 0$. On division by Δ which is not zero and as $\mathscr{C} \equiv J$ the non-homogeneous equation of the pair of asymptotes is

$$JS - \Delta = 0. \tag{75.1}$$

The gradients of the asymptotes are those of the lines parallel to them through the origin. These lines are $ax^2 + 2hxy + by^2 = 0$, and so the gradients γ of the asymptotes are the roots of the quadratic equation $a + 2h\gamma + b\gamma^2 = 0$. Comparing this with (73.3) we see that each asymptote is a self-conjugate diameter.

A rectangular hyperbola has orthogonal asymptotes. Consequently the necessary and sufficient condition that $S = 0$ be a rectangular hyperbola is that the invariant $I = 0$.

In the case of the central conic, it cuts the line at infinity at the points given by $ax^2 + 2hxy + by^2 = z = 0$. From (75.1) we see that the asymptotes also cut the line at infinity at these same points. Thus the asymptotes are the tangents to the conic at its points of infinity.

For a parabola $J = 0$ and (75.1) in homogeneous coordinates reduces to $z^2 = 0$. That is, we can regard the line at infinity as a double asymptote of a parabola.

Example 15. Obtain the equations of the asymptotes of the following conics:

(*a*) $3x^2 + 8xy - 3y^2 + 6x + 8y + 4 = 0,$

(*b*) $2x^2 + 9xy - 5y^2 + 2y - 7 = 0.$

76. Focus and directrix

The curve, defined as the locus of a point which moves so that its distance from a focus (α, β), is in a constant ratio e (called the eccentricity) to its distance from a directrix $lx + my + n = 0$, has the equation

$$(x-\alpha)^2 + (y-\beta)^2 - \frac{e^2}{l^2+m^2}(lx+my+n)^2 = 0. \quad (76.1)$$

This equation contains five effective constants, namely α, β, e and the ratios $l:m:n$. Consequently we can identify this locus, which is quadratic, with the conic $S = 0$. The problem arises: given the conic $S = 0$; to determine the foci (α, β) and the corresponding directrices.

The pair of complex conjugate straight lines through (α, β)

$$x - \alpha \pm i(y-\beta) = 0 \quad (76.2)$$

intersect the conic (76.1) where $(lx + my + n)^2 = 0$. That is, in coincident points. In homogeneous coordinates the selines are $x - \alpha z \pm i(y - \beta z) = 0$ and so we see that the circular points at infinity $I \equiv (1, i, 0)$ and $J \equiv (1, -i, 0)$ lie one on each of the straight lines. Thus the foci are to be found at the points of intersection of the two pairs of tangents from the circular points to a conic.

The polar of (α, β) with respect to the conic (76.1) is

$$\left\{\left(1 - \frac{e^2l^2}{l^2+m^2}\right)\alpha - \frac{e^2lm}{l^2+m^2}\beta - \left(\alpha + \frac{e^2ln}{l^2+m^2}\right)\right\}x$$

$$+\left\{-\frac{e^2lm}{l^2+m^2}\alpha + \left(1 - \frac{e^2m^2}{l^2+m^2}\right)\beta - \left(\beta + \frac{e^2mn}{l^2+m^2}\right)\right\}y$$

$$+\left\{-\left(\alpha + \frac{e^2ln}{l^2+m^2}\right)\alpha - \left(\beta + \frac{e^2mn}{l^2+m^2}\right)\beta\right.$$

$$\left. + \alpha^2 + \beta^2 - \frac{e^2n^2}{l^2+m^2}\right\} = 0,$$

which simplifies on multiplication by $l^2 + m^2$ and division by $e^2(l\alpha + m\beta + n)$ (which is not zero if we stipulate that the focus must not lie on the straight line chosen as directrix) to $lx + my + n = 0$. Hence the directrix is the polar of the corresponding focus.

The straight line $lx + my + n = 0$ is a tangent to $S = 0$ if condition (70.3) is satisfied. For one of the straight lines (69.2) we have $l = 1$, $m = i$ and $n = -(\alpha + i\beta)$ and so

$$\mathscr{A} - \mathscr{B} + \mathscr{C}(\alpha^2 - \beta^2 + 2i\alpha\beta) - 2(i\alpha - \beta)\mathscr{F}$$

$$- 2(\alpha + i\beta)\mathscr{G} + 2i\mathscr{H} = 0.$$

For the other tangent line of (69.2) we similarly obtain

$$\mathscr{A} - \mathscr{B} + \mathscr{C}(\alpha^2 - \beta^2 - 2i\alpha\beta) - 2(-i\alpha - \beta)\mathscr{F} \\ - 2(\alpha - i\beta)\mathscr{G} - 2i\mathscr{H} = 0.$$

These two equations are equivalent to the two real equations

$$\left.\begin{aligned} \mathscr{C}(\alpha^2 - \beta^2) - 2\mathscr{G}\alpha + 2\mathscr{F}\beta + \mathscr{A} - \mathscr{B} = 0, \\ \mathscr{C}\alpha\beta - \mathscr{F}\alpha - \mathscr{G}\beta + \mathscr{H} = 0. \end{aligned}\right\} \qquad (76.3)$$

The substitution† $\mathscr{C}\alpha - \mathscr{G} = \mathscr{C}\gamma$; $\mathscr{C}\beta - \mathscr{F} = \mathscr{C}\delta$ reduces these equations to

$$\mathscr{C}^2(\gamma^2 - \delta^2) = \mathscr{C}(\mathscr{B} - \mathscr{A}) + \mathscr{G}^2 - \mathscr{F}^2 = (a - b)\Delta,$$
$$\mathscr{C}^2\gamma\delta = \mathscr{G}\mathscr{F} - \mathscr{C}\mathscr{H} = h\Delta,$$

in virtue of (67.3). Eliminating δ we obtain the quartic equation

$$\mathscr{C}^4\gamma^4 - (a - b)\mathscr{C}^2\Delta\gamma^2 - h^2\Delta^2 = 0. \qquad (76.4)$$

This is quadratic in γ^2 and its discriminant is $(a - b)^2\mathscr{C}^4\Delta^2 + 4h^2\mathscr{C}^4\Delta^2$, which is positive, and so both roots in γ^2 are real. The product of the roots is $-h^2\Delta^2/\mathscr{C}^4$ and is therefore negative. Consequently one root in γ^2 is positive and the other negative. Thus the quartic has two real roots and two conjugate complex roots. That is, a conic has four foci, two of which are real.

The reader is advised not to memorize these results but rather to study the method and solve numerical problems from first principles.

If the conic $S = 0$ is a parabola, we have $\mathscr{C} \equiv J = 0$ and equations (76.3) become

$$\left.\begin{aligned} 2\mathscr{G}\alpha - 2\mathscr{F}\beta = \mathscr{A} - \mathscr{B}, \\ \mathscr{F}\alpha + \mathscr{G}\beta = \mathscr{H}. \end{aligned}\right\} \qquad (76.5)$$

These equations always have one solution, which shows that a parabola has one real focus at

$$(\mathscr{A}\mathscr{G} - \mathscr{B}\mathscr{G} + 2\mathscr{F}\mathscr{H},\ \ \mathscr{B}\mathscr{F} - \mathscr{A}\mathscr{F} + 2\mathscr{G}\mathscr{H},\ \ 2\mathscr{F}^2 + 2\mathscr{G}^2).$$

The corresponding directrix can be obtained by writing down the polar of the focus.

Example 16. Find the real foci and directrices of the conics whose equations are:

(a) $x^2 + 2xy + y^2 + 2x = 0$.

(b) $6x^2 + 4xy + 9y^2 + 28x - 14y + 41 = 0$.

(c) $x^2 + y^2 + (x \cos \theta + y \sin \theta)^2 = 1$.

---

† This substitution is merely a translation to parallel axes through the centre of the conic.

77. Eccentricity

We now calculate the eccentricity of the conic $S = 0$. Comparing coefficients with (76.1), we have

$$a = \mu\left\{1 - \frac{e^2l^2}{l^2 + m^2}\right\}; \quad h = -\frac{\mu e^2 lm}{l^2 + m^2}; \quad b = \mu\left\{1 - \frac{e^2m^2}{l^2 + m^2}\right\},$$

where μ is some factor of proportionality. Then the invariants

$$I \equiv a + b = \mu(2 - e^2),$$
$$J \equiv ab - h^2 = \mu^2(1 - e^2).$$

Thus e^2 satisfies the equation $(1 - e^2)I^2 = (2 - e^2)^2 J$. That is,

$$Je^4 + \{I^2 - 4J\}e^2 - \{I^2 - 4J\} = 0. \tag{77.1}$$

The discriminant of this quadratic equation in e^2 is

$$(I^2 - 4J)^2 + 4J(I^2 - 4J) = I^2(I^2 - 4J) = I^2[(a - b)^2 + 4h^2] > 0$$

and so both roots in e^2 are always real. Let the roots be $e_1{}^2$ and $e_2{}^2$. Then

$$\left.\begin{aligned} \frac{1}{e_1{}^2} + \frac{1}{e_2{}^2} &= 1, \\ e_1{}^2e_2{}^2 = -\frac{I^2 - 4J}{J} &= -\frac{(a - b)^2 + 4h^2}{J} \end{aligned}\right\} \tag{77.2}$$

If the conic is an ellipse $J > 0$ and so $e_1{}^2e_2{}^2 < 0$. Consequently one root (say) $e_1{}^2 > 0$ whilst the other $e_2{}^2 < 0$. Hence from (77.2) we see that $1/e_1{}^2 > 1$ and so $e_1 < 1$. Clearly the real eccentricity corresponds to the two real foci.

If the conic is a hyperbola $J < 0$ and so $e_1{}^2e_2{}^2 > 0$. From (77.2) it then follows that $e_1{}^2$ and $e_2{}^2$ are both positive. Neither $e_1{}^2 < 1$ nor $e_2{}^2 < 1$ because then (77.2) would yield $e_2{}^2 < 0$ and $e_1{}^2 < 0$ respectively. Thus $e_1{}^2 > 1$ and $e_2{}^2 > 1$. Accordingly a hyperbola has two real eccentricities. In order to see which value corresponds to the real focus it is necessary to test by means of (76.1).

If the conic is a parabola $J = 0$ and clearly $e = 1$.

Example 17. Calculate the eccentricities of the conics in Example 16.

Miscellaneous examples

1. Obtain the conditions that the line-pair $a'x^2 + 2h'xy + b'y^2 + 2g'x + 2f'y + c' = 0$ be conjugate with respect to the conic $ax^2 + 2hxy + by^2 + 2gx + 2fy + c = 0$.

2. Show that the angle α between the pair of tangents from A to the conic $S = 0$ is given by $\tan^2 \alpha = -4\Delta S_A/\Theta_A{}^2$, where $\Theta_A \equiv \mathscr{C}(x_A{}^2 + y_A{}^2) - 2\mathscr{G}x_A - 2\mathscr{F}y_A + \mathscr{A} + \mathscr{B}$.

Hence show that the locus of the points of intersection of tangents which make a constant angle α with one another is the curve of 4th degree $\Theta^2 \tan^2 \alpha + 4\Delta S = 0$. (This curve is called the **isoptic locus.**)

3. Prove that the isoptic locus of a parabola is in general a conic. Further show that the standard form of this conic is $x^2 - y^2 \tan^2 \alpha = \Delta \operatorname{cosec}^2 \alpha / I^3$ $(\alpha \neq \pi/2)$.

4. Show that the centre of the isoptic locus of a parabola always lies on its axis.

5. Obtain the real foci and the eccentricities of the hyperbola $x^2 + 2hxy + y^2 = 1$ $(h > 1)$.

6. Find the equations of the directrix and axis of the parabola $(\lambda x + \mu y)^2 = 2\rho x$. Further obtain the coordinates of the focus.

7. Prove that the straight line $lx + my + n = 0$ is a principal axis of the conic $S = 0$ if $\dfrac{al + hm}{l} = \dfrac{hl + bm}{m} = \dfrac{gl + fm}{n}$. Deduce that the equation of the principal axes is $h\left\{\left(\dfrac{\partial S}{\partial x}\right)^2 - \left(\dfrac{\partial S}{\partial y}\right)^2\right\} - (a - b)\dfrac{\partial S}{\partial x}\dfrac{\partial S}{\partial y} = 0.$

8. Show that the equation of a conic referred to the tangent and normal at a point on it as axes is $ax^2 + 2hxy + by^2 + 2fy = 0$. Further, prove that the gradients of the three normals from the origin to the conic are the roots of the cubic equation $2hg^3 + (2a - b)g^2 + a = 0$.

9. Obtain the necessary and sufficient conditions that the pair of straight lines $\lambda x^2 + 2\mu xy + \nu y^2 = 0$ be conjugate diameters with respect to the conic $ax^2 + 2hxy + by^2 + c = 0$. Hence, or otherwise, find the equation of the two diameters which are conjugate with respect to both the conics $2x^2 + 4xy - 3y^2 = 1$ and $6x^2 - 10xy + 7y^2 = 1$.

10. Show that the equations $x = at^2 + 2bt + c$; $y = a't^2 + 2b't + c'$ define a parabola. Calculate the length of its latus rectum.

11. If $S = 0$ represents a hyperbola, show that the conjugate hyperbola is $S - 2\Delta/J = 0$.

Answers

1. $3x^2 - y^2 - 3x + y = 0$. **2.** Two; $(x - y)^2 - 8(x + y) = 0$, $(5x + 3y)^2 - 8(25x + 9y) = 0$. **4.** (*a*) $(1, 0)$ and $(2, 2)$; (*b*) $(2, 2)$; (*c*) $(1, 0)$ and $(12/5, 7/5)$. **5.** $2x - 3y = 0$, $16x + 37y - 78 = 0$, $36x + 7y - 78 = 0$. **6.** $(-1, 1)$, (*a*) no, (*b*) no, (*c*) yes, (*d*) yes. **9.** $\tan^{-1}(\sqrt{33}/20)$. **11.** $12x + 4y = 3$. **12.** (*a*) $(-1, 0)$; (*b*) $(2, 3)$; (*c*) $(-1/2, -2)$. **13.** $16x - 8y + 7 = 0$. **14.** (*a*) $x - 2y + 1 = 0$, $2x + y + 2 = 0$; (*b*) $x - y + 1 = 0$; $x + y - 5 = 0$; (*c*) $(1 \pm \sqrt{5})(2x + 1) + 4(y + 2) = 0$. **15.** (*a*) $3x - y + 3 = 0$, $x + 3y + 1 = 0$; (*b*) $11x + 55y - 2 = 0$, $22x - 11y + 4 = 0$. **16.** (*a*) $(-1/4, -1/4)$, $2x - 2y - 1 = 0$; (*b*) $(-2, 1)$, $2x - y + 3 = 0$; $(-18/5, 9/5)$, $2x - y + 11 = 0$; (*c*) $(-\sin\theta/\sqrt{2}, \cos\theta/\sqrt{2})$, $x\sin\theta - y\cos\theta + \sqrt{2} = 0$; $(\sin\theta/\sqrt{2}, -\cos\theta/\sqrt{2})$, $x\sin\theta - y\cos\theta - \sqrt{2} = 0$. **17.** (*a*) 1; (*b*) $1/\sqrt{2}$; (*c*) $1/\sqrt{2}$.

Miscellaneous examples. 1. $\Delta' = 0$ and $a'\mathscr{A} + b'\mathscr{B} + c'\mathscr{C} + 2f'\mathscr{F} + 2g'\mathscr{G} + 2h'\mathscr{H} = 0$. **5.** $(\pm\sqrt{h/(h^2 - 1)}, \pm\sqrt{h/(h^2 - 1)})$, $2h/(h - 1)$ and $2h/(h + 1)$. **6.** $2\mu(\mu x - \lambda y) + \rho = 0$, $(\lambda^2 + \mu^2)(\lambda x + \mu y) - \rho\lambda = 0$, $(\rho/2(\lambda^2 + \mu^2), \rho\lambda/2\mu(\lambda^2 + \mu^2))$. **9.** $a\nu - 2h\mu + b\lambda = 0$, $22x^2 - 32xy + y^2 = 0$. **10.** $2(b'a - a'b)^2/(a^2 + a'^2)^{3/2}$.

CHAPTER XI

PENCIL OF CONICS

78. Pencil of conics

Consider the two conics

$$S \equiv ax^2 + 2hxy + by^2 + 2gx + 2fy + c = 0,$$
$$S' \equiv a'x^2 + 2h'xy + b'y^2 + 2g'x + 2f'y + c' = 0,$$

and set up the equation

$$S + \lambda S' = 0. \tag{78.1}$$

This equation is of the second degree in x and y and thus represents a conic, which must pass through the common points of the conics S and S' given by $S = S' = 0$. These two equations are of the second degree and so in general S and S' have four points of intersection. Consequently $S + \lambda S' = 0$ represents a system of conics passing through four given points called **base points.** Such a system is called a **pencil** of conics.

Through any point, not a base point, there passes one and only one conic of the pencil because (78.1) yields a linear equation in λ.

Two conics of a pencil touch a given straight line because the tangency condition (70.3) is quadratic in λ.

The equation of a pencil of conics through the four given points A, B, C, D is

$$u_1u_2 + \lambda u_3u_4 = 0,$$

where $u_1 = 0$, $u_2 = 0$, $u_3 = 0$, $u_4 = 0$ are the equations of the straight lines AB, CD, AD, BC respectively.

The conic $S + \lambda S' = 0$ is a rectangular hyperbola if

$$a + b + \lambda(a' + b') = 0;$$

hence a pencil of conics contains one rectangular hyperbola. (All the conics of the pencil are rectangular hyperbolas if this equation is an identity.)

The condition that $S + \lambda S' = 0$ be a parabola is

$$(a + \lambda a')(b + \lambda b') - (h + \lambda h')^2 = 0$$

and so in general a pencil has two parabolas. Again an identity would mean that all conics of the pencil are parabolas.

The conic $S + \lambda S' = 0$ degenerates to a line-pair if

$$\begin{vmatrix} a + \lambda a' & h + \lambda h' & g + \lambda g' \\ h + \lambda h' & b + \lambda b' & f + \lambda f' \\ g + \lambda g' & f + \lambda f' & c + \lambda c' \end{vmatrix} = 0.$$

This is a cubic equation in λ. Accordingly there are three line-pairs in a pencil of conics.

Example 1. Show that the polar of a fixed point with respect to any conic of a pencil passes through a fixed point.

Example 2. Show that the locus of the centres of the conics of a pencil is a conic if the base points are distinct.

Example 3. If two conics of a pencil are rectangular hyperbolas, show that all conics of the pencil are rectangular hyperbolas.

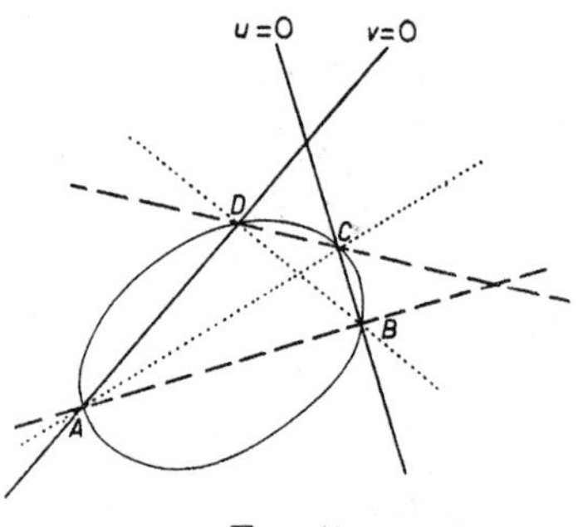

FIG. 41

79. Contact of conics

If S' degenerates to the line-pair $uv = 0$, then the pencil $S + \lambda uv = 0$ represents the pencil of conics through the four points of intersection A, B, C and D of the conic $S = 0$ and the two straight lines $uv = 0$. The other two degenerate line-pairs of the pencil are shown in Fig. 41.

However, the straight lines $u = 0$ and $v = 0$ may bear some special relation to the conic $S = 0$ and we proceed to investigate the various possibilities.

(1) Let A and D coincide, as in Fig. 42. Then all conics of the pencil have coincident points at A, that is they have a common tangent there. The conics are said to have **single-contact** at A and the pencil can be represented by the equation

$$S + \lambda T_A v = 0,$$

where $T_A = 0$ is the common tangent of the pencil at the point A.

The degenerate conics of this pencil are (*a*) the tangent at A with the chord BC, (*b*) the line-pair AC, AB taken twice.

(2) Let the pairs of points A and D, B and C coincide as in Fig. 43. Then all conics of the pencil have common tangents at A and B. We say that the conics of the pencil have **double-contact.** The pencil is represented by

$$S + \lambda T_A T_B = 0.$$

The degenerate conics of the pencil are (*a*) the tangents at A and B, (*b*) the repeated chord AB taken twice.

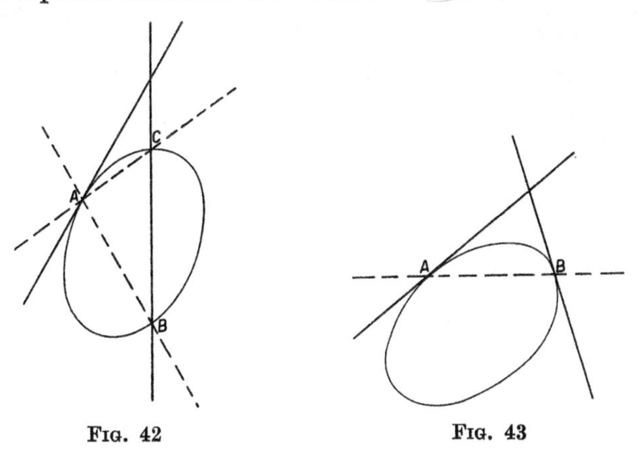

FIG. 42 FIG. 43

A pencil of conics having double contact with $S = 0$ at the points A and B is also represented by $S + \lambda u^2 = 0$, where $u = 0$ is the equation of the line AB.

(3) Let the points A, C and D coincide as in Fig. 44. Then all conics of the pencil have three coincident points at A, and we say that there is **three-point-contact.** If the equation of the chord AB is

$$l(x - x_A) + m(y - y_A) = 0,$$

then the pencil of conics is represented by

$$S + \lambda T_A\{l(x - x_A) + m(y - y_A)\} = 0.$$

The degenerate conics of this pencil are the tangent at A and the chord AB, counted three times.

(4) Finally suppose the four points A, B, C and D coincide. Then we say that all conics of the pencil have **four-point contact** at A and they are represented by the equation

$$S + \lambda T_A^2 = 0.$$

The degenerate conics of this pencil are the coincident tangent lines at A counted three times.

Example 4. If a circle has double contact with a conic, show that the chord of contact is perpendicular to an axis.

Example 5. Show that the locus of the centres of the conics which have four-point contact with a given conic at a given point is a straight line through the given point.

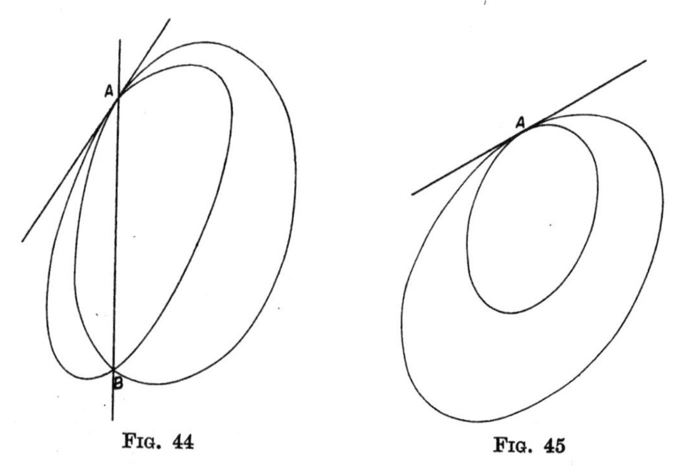

FIG. 44 FIG. 45

80. Circle of curvature

The pencil of conics having three-point-contact at A with the conic $S = 0$ is given by the equation

$$\begin{aligned} ax^2 + 2hxy + by^2 + 2gx + 2fy + c \\ + \lambda\{(ax_A + hy_A + g)x + (hx_A + by_A + f)y \\ + gx_A + fy_A + c\}\{l(x - x_A) + m(y - y_A)\} = 0. \quad (80.1)\end{aligned}$$

Are there any circles in this pencil? The necessary and sufficient conditions that (80.1) be a circle are

$$a + \lambda l(ax_A + hy_A + g) = b + \lambda m(hx_A + by_A + f),$$
$$2h + \lambda\{m(ax_A + hy_A + g) + l(hx_A + by_A + f)\} = 0.$$

We have two linear equations for λl and λm; hence there is one solution. We substitute this solution in (80.1) and the circle so obtained is called the **circle of curvature** or **osculating circle** at A and its centre, the **centre of curvature**† at A.

† The reader is asked to reconcile this definition with the results of Exercises 8, p. 44, 9, page 53 and 5, page 57.

The locus of the centre of curvature corresponding to all points on a curve is called its **evolute.**

Suppose we tackle the problem: "How many circles have four-point-contact at A with the conic $S = 0$?" That is, the equation

$$ax^2 + 2hxy + by^2 + 2gx + 2fy + c + \lambda\{(ax_A + hy_A + g)x + (hx_A + by_A + f)y + gx_A + fy_A + c\}^2 = 0 \quad (80.2)$$

represents a circle. Therefore,

$$a + \lambda(ax_A + hy_A + g)^2 = b + \lambda(hx_A + by_A + f)^2,$$
$$h + \lambda(ax_A + hy_A + g)(hx_A + by_A + f) = 0.$$

Two equations in one unknown are in general inconsistent. Eliminating λ we have

$$\frac{(ax_A + hy_A + g)^2 - (hx_A + by_A + f)^2}{a - b} = \frac{(ax_A + hy_A + g)(hx_A + by_A + f)}{h}.$$

The reader is asked to verify that this equation is satisfied if the point A lies on the principal axes (74.2) of the conic. Accordingly the circle of curvature has four-point-contact with a conic at a vertex only.

If equation (80.2) represents a parabola or rectangular hyperbola, we refer to it as the **osculating parabola** and **osculating rectangular hyperbola** respectively.

Example 6. Obtain the equation of the circle of curvature at the point (2, 1) on the ellipse $x^2 + 2y^2 = 6$.

Example 7. Show that the locus of the centres of osculating rectangular hyperbolas of the parabola $y^2 = 4ax$ is an equal parabola.

Example 8. Find the equation of the osculating parabola of the rectangular hyperbola $x^2 - y^2 = a^2$ at the point $(a \sec \theta, a \tan \theta)$.

81. Pair of tangents at ends of a chord

The pencil of conics having double contact with $S = 0$ at the points A and B (Fig. 43) is given by $S + \lambda T_A T_B = 0$. The chord AB taken twice is also a degenerate member of the pencil and so the pencil can also be represented by

$$S + \lambda(lx + my + n)^2 = 0,$$

where $lx + my + n = 0$ is the equation of the chord AB. The pair of tangents at A and B correspond to the other degenerate conic of the pencil, for which

$$\begin{vmatrix} a + \lambda l^2 & h + \lambda lm & g + \lambda ln \\ h + \lambda lm & b + \lambda m^2 & f + \lambda mn \\ g + \lambda ln & f + \lambda mn & c + \lambda n^2 \end{vmatrix} = 0.$$

This equation reduces to

$$\Delta + \lambda(\mathscr{A}l^2 + \mathscr{B}m^2 + \mathscr{C}n^2 + 2\mathscr{F}mn + 2\mathscr{G}nl + 2\mathscr{H}lm) = 0.$$

Consequently the pair of tangents to the conic $S = 0$ at the ends of the chord $lx + my + n = 0$ is given by

$$(\mathscr{A}l^2 + \mathscr{B}m^2 + \mathscr{C}n^2 + 2\mathscr{F}mn + 2\mathscr{G}nl + 2\mathscr{H}lm)S - \Delta(lx + my + n)^2 = 0.$$

Miscellaneous examples

1. Show that the orthoptic loci of the pencil of conics which touch the coordinate axes at two fixed points form a coaxal system of circles.

2. Prove that the locus of the centres of conics which pass through the vertices and the orthocentre of a triangle is the nine-points circle of the triangle.

3. Obtain the equation of (a) the osculating circle, (b) the osculating parabola, (c) the osculating rectangular hyperbola at the origin of the conic $ax^2 + 2hxy + by^2 + 2gx = 0$.

4. The polars of A and B with respect to a conic, intersect it in the points P, Q, R and S. Show that there is a conic passing through these six points.

5. Show that the locus of the poles of a given straight line with respect to the conics of a pencil is a conic.

6. If $\bar{S} = 0$ is the conic which is the locus of the centres of all conics through four given distinct points, show that (a) the principal axes of $\bar{S} = 0$ are parallel to the asymptotes of the rectangular hyperbola through the four points, (b) the polars of any fixed point on $\bar{S} = 0$ with respect to the conics of the pencil are parallel lines.

7. A pencil of conics passes through the vertices of a rhombus. Show that their centres must all be at the point of intersection of the diagonals. Prove that this result is also true for a parallelogram.

8. Show that the four points of intersection of two central conics whose axes are parallel are generally concyclic.

9. Two conics S_1 and S_2 have double contact. Two straight lines are drawn through a point on their chord of contact, cutting S_1 in four points. Show that it is possible to draw another conic through these four points to have double contact with S_2.

Answers

6. $3(x^2 + y^2) - 4x + 2y - 9 = 0$. **7.** $y^2 = -4a(x + 2a)$. **8.** $(x \tan\theta - y \sec\theta)^2 - 2a(x \sec\theta - y \tan\theta) = 0$.

Miscellaneous examples. **3.** (a) $b(x^2 + y^2) + 2gx = 0$, (b) $(hx + by)^2 + 2bgx = 0$, (c) $b(x^2 - y^2) - 2hxy - 2gx = 0$.

CHAPTER XII

HOMOGRAPHIC CORRESPONDENCE

82. Cross-ratio

We define the **cross-ratio** of the four collinear points A, B, C and D (Fig. 46) to be

$$\mu \equiv \{A, B; C, D\} = \frac{AC}{CB} \Big/ \frac{AD}{DB}.$$

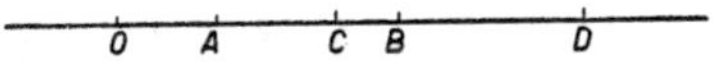

Fig. 46

There are twenty-four arrangements of the order of the letters $ABCD$ but there are only six different values of the cross-ratio, since

$$\{A, B; C, D\} = \{B, A; D, C\} = \{C, D; A, B\} = \{D, C; B, A\}.$$

Select a fixed point O as origin on the straight line $ABCD$ and denote the distances OA, OB, OC and OD by α, β, γ and δ respectively. Then

$$\mu = -\frac{(\gamma - \alpha)(\beta - \delta)}{(\beta - \gamma)(\alpha - \delta)},$$

and we also say that μ is the cross-ratio $\{\alpha, \beta; \gamma, \delta\}$ of the four numbers α, β, γ and δ.

Introduce the three products

$$p = (\beta - \gamma)(\alpha - \delta); \quad q = (\gamma - \alpha)(\beta - \delta); \quad r = (\alpha - \beta)(\gamma - \delta)$$

which satisfy the identity

$$p + q + r = 0.$$

Then the six cross-ratios of the four numbers α, β, γ, δ taken in various orders are

$$-\frac{q}{p}, \quad -\frac{p}{q}, \quad -\frac{r}{p}, \quad -\frac{p}{r}, \quad -\frac{r}{q}, \quad -\frac{q}{r}.$$

We choose $\mu = \{A, B; C, D\} = -p/q$ and we readily verify that the six values of the cross-ratios are

$$\mu, \quad \frac{1}{\mu}, \quad 1 - \mu, \quad \frac{1}{1 - \mu}, \quad \frac{\mu - 1}{\mu}, \quad \frac{\mu}{\mu - 1}.$$

If two of α, β, γ, δ coincide, then one of the numbers p, q or r is zero and so the six cross-ratios are $0, \infty, 1, 1, \infty, 0$.

If three of α, β, γ, δ coincide, then $p = q = r = 0$ and all the cross-ratios are indeterminate.

Example 1. Calculate the following cross-ratios: (*a*) $\{1, 2;\ 3, 4\}$, (*b*) $\left\{1, \dfrac{1+\lambda}{1-\lambda};\ \lambda, \dfrac{1-\lambda}{1+\lambda}\right\}$, (*c*) $\{1, \infty;\ \sqrt{2}, -1\}$, (*d*) $\{1, -1;\ \omega, \omega^2\}$, where ω is a complex cube root of unity.

83. Harmonic Case

We now investigate the possibility that two cross-ratios of four distinct points be equal. The quantities p. q and r defined by (82.1) must be non-zero if the four points are distinct. Let one of the cross-ratios be $\mu = -p/q$ then there are five possibilities.

(i) $\mu = 1/\mu$, from which $\mu = 1$ or $\mu = -1$. If $\mu = 1$, $r = 0$ and so this value is excluded. The other value $\mu = -1$ yields the six values of the cross-ratios to be $-1, -1, 2, \frac{1}{2}, 2, \frac{1}{2}$. This cross-ratio is called **harmonic.**

(ii) $\mu = 1 - \mu$ from which $\mu = \frac{1}{2}$ and this yields the harmonic case.

(iii) $\mu = 1/(1 - \mu)$, from which $\mu^2 - \mu + 1 = 0$ and so $\mu = -\omega$ or $-\omega^2$, where ω is a complex cube root of $+1$ (that is, $\omega^3 = +1$, but $\omega \neq +1$). The six values of the cross-ratios are $-\omega, -\omega^2, -\omega^2, -\omega, -\omega, -\omega^2$ and we call this the **equianharmonic** case. It cannot occur if the numbers α, β, γ and δ are real.

(iv) $\mu = (\mu - 1)/\mu$ and so $\mu^2 - \mu + 1 = 0$ and once more we have the equianharmonic case.

(v) $\mu = \mu/(\mu - 1)$, from which $\mu = 0$ or $\mu = 2$. The case $\mu = 0$ is excluded as it makes $p = 0$ and $\mu = 2$ merely repeats the harmonic case.

In particular note that the cross-ratio $\{0, \infty;\ 1, -1\} = -1$.

Example 2. If A and B divide C and D harmonically, show that the cross-ratios $\{A, B;\ C, D\} = -1$.

84. Homographic correspondence

A **homographic correspondence** (or one-one algebraic correspondence) between two variables x and y is defined by

$$axy + bx + cy + d = 0. \tag{84.1}$$

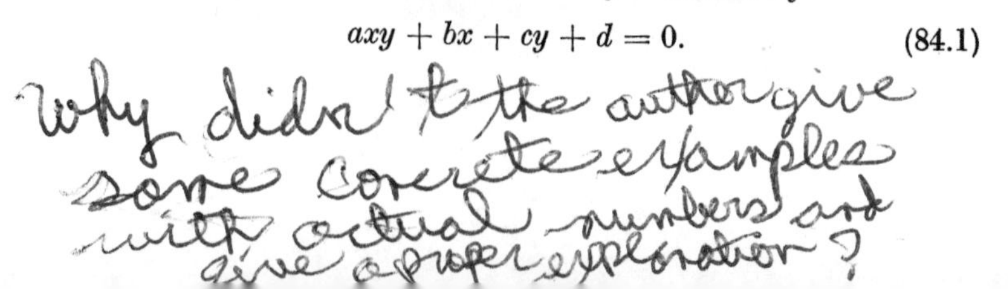

This is a bilinear equation, which can be written

$$(ax + c)(ay + b) + ad - bc = 0.$$

It will always be assumed that $ad - bc \neq 0$ and so we cannot factorize the expression $axy + bx + cy + d$. Consequently to each value of x or y there corresponds exactly one value of y or x respectively.

Note very carefully that if $y = \gamma$ when $x = \delta$, it does not follow that in general $y = \delta$ when $x = \gamma$.

A homographic correspondence involves three effective constants, namely the ratios $a:b:c:d$. Thus a homographic correspondence is uniquely determined by three given pairs of numbers.

α is called a **double point** if $y = \alpha$ when $x = \alpha$. Hence α satisfies the quadratic equation

$$at^2 + (b + c)t + d = 0.$$

This equation has two roots and so there are two double points and the homographic correspondence is said to be

$$\left\{\begin{array}{l}\text{hyperbolic}\\ \text{parabolic}\\ \text{elliptic}\end{array}\right\}\begin{array}{l}\text{according}\\ \text{as the double}\\ \text{points are}\end{array}\left\{\begin{array}{l}\text{real and distinct}\\ \text{coincident}\\ \text{conjugate complex}\end{array}\right\}\begin{array}{c}\text{That is}\\ (q + r)^2 - 4ps\\ \end{array}\left\{\begin{array}{c}>\\ =\\ <\end{array}\right\}0.$$

If one double point is at infinity $a = 0$ and the homographic correspondence is

$$bx + cy + d = 0.$$

If the double points coincide at infinity $a = 0$, $b + c = 0$ and the homographic correspondence is

$$x - y = \text{constant}.$$

Example 3. State if x and y are connected by a homographic correspondence in the following cases. If so, calculate the values of the double points. (*a*) $x = e^t$, $y = e^{-t}$; (*b*) $x = (1 - t^2)^2/(1 + t^2)^2$, $y = 4t^2/(1 + t^2)^2$; (*c*) $x = \log t$, $y = \sin t$; (*d*) $x = \log at$, $y = \log bt$; (*e*) $x = 1/(1 + t)$, $y = 1/(1 - t)$; (*f*) $x = t^2$, $y = t^3$.

85. Fundamental cross-ratio theorem

Occasionally it is more convenient to define the homographic correspondence by

$$y = \frac{ax + b}{cx + d}; \quad ad - bc \neq 0.$$

This is equivalent to (84.1).

Consider the four numbers $\alpha_x, \beta_x, \gamma_x, \delta_x$, together with the corresponding four numbers $\alpha_y, \beta_y, \gamma_y, \delta_y$, where $\alpha_y = (a\alpha_x + b)/(c\alpha_x + d)$, etc. Then

$$\beta_y - \gamma_y = \frac{a\beta_x + b}{c\beta_x + d} - \frac{a\gamma_x + b}{c\gamma_x + d} = \frac{(ad - bc)(\beta_x - \gamma_x)}{(c\beta_x + d)(c\gamma_x + d)}.$$

Thus, with a notation similar to (82.1) we have

$$p_y \equiv (\beta_y - \gamma_y)(\alpha_y - \delta_y) = \frac{(ad - bc)^2 p_x}{(c\alpha_x + d)(c\beta_x + d)(c\gamma_x + d)(c\delta_x + d)}.$$

But $ad - bc \neq 0$ and so

$$p_y : q_y : r_y = p_x : q_x : r_x.$$

Hence the cross-ratios of the four numbers $\alpha_y, \beta_y, \gamma_y, \delta_y$ are equal to the cross-ratios of $\alpha_x, \beta_x, \gamma_x, \delta_x$. That is, we have proved the important result that *cross-ratio is unaltered by a homographic correspondence.*

We can now easily set up the homographic correspondence in which the three numbers x_1, x_2, x_3 correspond to y_1, y_2, y_3. For, by the above theorem $\{xx_1;\ x_2x_3\} = \{yy_1;\ y_2y_3\}$ and so the correspondence is

$$\frac{(x - x_2)(x_1 - x_3)}{(x - x_3)(x_1 - x_2)} = \frac{(y - y_2)(y_1 - y_3)}{(y - y_3)(y_1 - y_2)}.$$

Example 4. Obtain the homographic correspondence between x and y in which 1 and 2 are double points and $x = 3$ corresponds to $y = 4$.

86. Involution

An **involution** is a symmetrical homographic correspondence. Thus an involution exists between the two variables x and y if

$$axy + b(x + y) + d = 0. \tag{86.1}$$

subject to the condition $ad - b^2 \neq 0$.

An involution involves two effective constants, namely the ratios $a:b:d$. Thus an involution is uniquely determined by two given pairs of numbers.

The double points α satisfy the quadratic equation

$$at^2 + 2bt + d = 0$$

and cannot coincide since $ad - b^2 \neq 0$. The involution is said to be

$\left\{\begin{matrix}\text{hyperbolic}\\ \text{elliptic}\end{matrix}\right\}$ according as the double points are $\left\{\begin{matrix}\text{real}\\ \text{conjugate complex}\end{matrix}\right\}$ That is $ad - b^2 \left\{\begin{matrix}<\\ >\end{matrix}\right\} 0.$

If one double point is at infinity, we have $a = 0$ and the involution is given by

$$x + y = \text{constant}.$$

Consider now the involution which has the given double points α and β. The homographic correspondence between x and y must be the bilinear equation

$$(x - \alpha)(y - \beta) + k(x - \beta)(y - \alpha) = 0. \tag{86.2}$$

For an involution, this relation must be symmetrical in x and y. That is, $-\beta - \alpha k = -\alpha - k\beta$, from which $k = +1$. Thus

$$\frac{(x - \alpha)(y - \beta)}{(x - \beta)(y - \alpha)} = -1.$$

Hence we deduce that any pair of points of an involution divide the double points harmonically.

Further from (86.2) we see that any pair of points of a homographic correspondence divide the double points (provided that they are finite and not coincident) in a constant cross-ratio. Note that (86.2) no longer represents the homographic correspondence if the double points coincide.

Example 5. State which of the homographic correspondences of Example 3 are involutions.

Example 6. Obtain the involution in which 1 is a double point and 2 corresponds to 3. Calculate the other double point.

87. Range of points

The position of a point P on the straight line AB is uniquely determined by the value of the ratio $\lambda = AP/PB$. By Joachimsthal's section formula the coordinates of P are

$$\left(\frac{x_A + \lambda x_B}{1 + \lambda}, \frac{y_A + \lambda y_B}{1 + \lambda}\right). \tag{87.1}$$

All points of the extended straight line are covered as λ varies from $-\infty$ to $+\infty$, and the range of values of λ is depicted in Fig. 47. The value $\lambda = -1$ corresponds to the point at infinity.

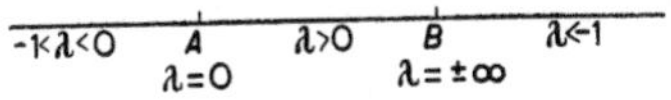

$-1<\lambda<0$ A $\lambda>0$ B $\lambda<-1$

$\lambda=0$ $\lambda=\pm\infty$

FIG. 47

A system of collinear points is called a **range** of points. If we specify the points of the range by Joachimsthal's section formula (87.1), A and B are called **base points** of the range.

Consider the four points P_1, P_2, P_3, P_4 of a range corresponding to the four numbers λ_1, λ_2, λ_3, λ_4 respectively. Let P_iL_i (Fig. 48) be drawn perpendicular to and $P_1M_2M_3M_4$ parallel to the x-axis. We have $P_1P_2:P_2P_3:P_3P_4 = P_1M_2:M_2M_3:M_3M_4$ and so

$$\{P_1, P_2;\ P_3, P_4\}$$
$$= \{P_1, M_2;\ M_3, M_4\} = \{L_1, L_2;\ L_3, L_4\},$$
$$= \left\{\frac{x_A + \lambda_1 x_B}{1 + \lambda_1}, \frac{x_A + \lambda_2 x_B}{1 + \lambda_2};\ \frac{x_A + \lambda_3 x_B}{1 + \lambda_3}, \frac{x_A + \lambda_4 x_B}{1 + \lambda_4}\right\}.$$

There exists a homographic correspondence between λ and $(x_A + \lambda x_B)/(1 + \lambda)$, and this homographic correspondence preserves

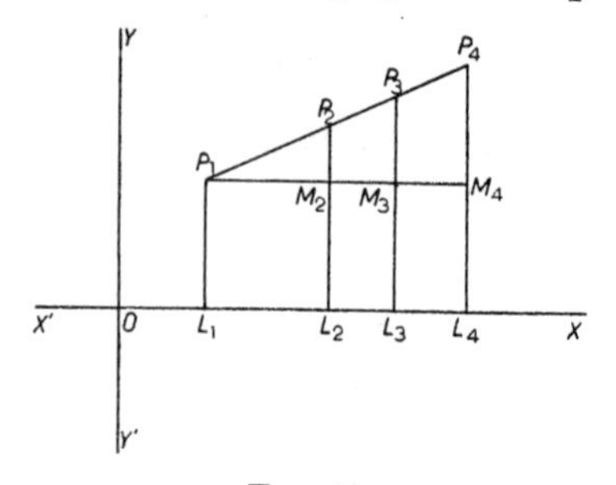

FIG. 48

cross-ratios (§85). Thus the cross-ratio of the four numbers $(x_A + \lambda_i x_B)/(1 + \lambda_i)$ equals the cross-ratio of the four numbers λ_i. Hence

$$\{P_1, P_2;\ P_3, P_4\} = \{\lambda_1, \lambda_2;\ \lambda_3, \lambda_4\}.$$

In particular, if the four points P_1, P_2, P_3, P_4 form a harmonic range, then $\{\lambda_1, \lambda_2;\ \lambda_3, \lambda_4\} = -1$.

Consequently the four points with homogeneous coordinates (x_A, y_A, z_A); (x_B, y_B, z_B); $(x_A + \lambda x_B, y_A + \lambda y_B, z_A + \lambda z_B)$; $(x_A - \lambda x_B, y_A - \lambda y_B, z_A - \lambda z_B)$ form a harmonic range.

Example 7. Choosing the base points of a range as $A \equiv (1, 2)$ and $B \equiv (5, 6)$, calculate the values of $\lambda \equiv AP_i/P_iB$ for the points $P_1 \equiv (7/3, 10/3)$, $P_2 \equiv (2, 3)$, $P_3 \equiv (11/3, 14/3)$ and $P_4 \equiv (-3, -2)$. Hence obtain the value of $\{P_1, P_2;\ P_3, P_4\}$.

88. Pencil of lines

A system of concurrent straight lines is called a **pencil** of straight lines, and the point of concurrency the **vertex.** Select two straight lines of the pencil, called base lines, with the equations

$$l_a x + m_a y + n_a z = 0 \quad \text{and} \quad l_b x + m_b y + n_b z = 0.$$

Then all straight lines of the pencil are represented by

$$l_a x + m_a y + n_a z + \lambda(l_b x + m_b y + n_b z) = 0, \qquad (88.1)$$

where λ varies from $-\infty$ to $+\infty$.

Consider the four lines l_1, l_2, l_3, l_4 of a pencil corresponding to $\lambda_1, \lambda_2, \lambda_3, \lambda_4$ respectively. These lines will intersect a line l (Fig. 49) not belonging to the pencil in four points L_1, L_2, L_3, L_4 respectively. Corresponding to each point L_i of l there is a unique line l_i of the pencil and corresponding to each line l_i of the pencil, there is a

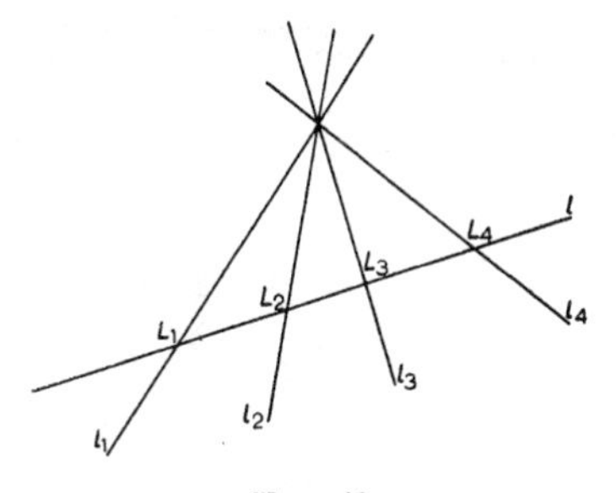

FIG. 49

unique point L_i on the line. Thus there is a one-one and clearly algebraic correspondence between the points L_i of the line l and the lines l_i of the pencil. That is, there is a homographic correspondence between the parameters λ_i which determine the line of the pencil and some parameter which determines the points of the range L_i on l. Thus the cross-ratio of the four points L_i is the cross-ratio of the four values of λ_i. This cross ratio is independent of the choice of the line l. Accordingly any straight line cuts four given lines of a pencil in a constant cross-ratio, called the cross-ratio of the four concurrent straight lines. We see that the cross-ratio of the four straight lines

$$l_a x + m_a y + n_a z + \lambda_i(l_b x + m_b y + n_b z) = 0; \quad (i = 1, 2, 3, 4)$$

is the cross-ratio of the four numbers λ_i.

If this cross ratio is -1, we say that the four straight lines form a **harmonic pencil.** We deduce that the four straight lines

$$l_a x + m_a y + n_a z = 0; \; l_b x + m_b y + n_b z = 0;$$
$$l_a x + m_a y + n_a z \pm \lambda(l_b x + m_b y + n_b z) = 0$$

form a harmonic pencil.

Miscellaneous examples

1. Show that the cross-ratio in which the roots of the two equations $a_1t^2 + 2h_1t + b_1 = 0$ and $a_2t^2 + 2h_2t + b_2 = 0$ divide each other is $(J_{12} + \sqrt{J_1J_2})/(J_{12} - \sqrt{J_1J_2})$ or its reciprocal, where $J_1 \equiv a_1b_1 - h_1^2$, $J_2 \equiv a_2b_2 - h_2^2$ and $2J_{12} = a_1b_2 + a_2b_1 - 2h_1h_2$. Hence deduce that the roots of the two quadratic equations separate each other harmonically if $J_{12} = 0$.

2. If the double points α and β of a homographic correspondence are distinct, show that its equation can be written

$$\frac{(x - \alpha)(y - \beta)}{(x - \beta)(y - \alpha)} = \text{constant.}$$

That is, the double points divide any pair of points of the homographic correspondence in a constant cross-ratio.

3. If a homographic correspondence has coincident double points α, show that its equation can be written

$$\frac{1}{x - \alpha} - \frac{1}{y - \alpha} = \text{constant.}$$

4. With the notation $Q_i \equiv a_it^2 + 2h_it + b_i$, show that the roots of the pencil of quadratic equations $Q_1 + \lambda Q_2 = 0$ are in involution.

5. Obtain the necessary and sufficient condition that the roots of $Q_1 = 0$, $Q_2 = 0$ and $Q_3 = 0$ are in involution.

6. Find the double points of the involution determined by the pencil of quadratic equations $Q_1 + \lambda Q_2 = 0$.

7. Prove that the four normals from the origin to the conic $ax^2 + 2hxy + by^2 + 2fy = 0$ form a harmonic pencil if $(2a - b)^3 + 54ah^2 = 0$. (Compare Example 8, page 90.)

8. If $P_i \equiv (at_i^2, 2at_i)$ for $i = 1, 2, 3, 4$ are four points on the parabola $y^2 = 4ax$, show that P_1P_2 and P_3P_4 are conjugate lines if $\{P_1P_2;\ P_3P_4\} = -1$.

9. Prove the corresponding result to Example 8 for the four points $P_i \equiv (ct_i, c/t_i)$ on the rectangular hyperbola $xy = c^2$.

Answers

1. (*a*) 4/3, (*b*) $2(1 - \lambda)/(1 + \lambda^2)$, (*c*) $(1 - \sqrt{2})/2$, (*d*) -1. **3.** (*a*) yes; ± 1; (*b*) yes; $\infty, \frac{1}{2}$; (*c*) no; (*d*) yes; ∞, ∞; (*e*) yes; 0, 1; (*f*) no. **4.** $xy + 2x - 5y + 2 = 0$. **5.** (*a*), (*b*) and (*e*). **6.** $3xy - 5(x + y) + 7 = 0$, 7/3. **7.** $\frac{1}{2}$, $\frac{1}{3}$, 2, $-\frac{1}{2}$; $\frac{3}{4}$.

Miscellaneous examples

5. $\begin{vmatrix} a_1 & h_1 & b_1 \\ a_2 & h_2 & b_2 \\ a_3 & h_3 & b_3 \end{vmatrix} = 0.$ **6.** Roots of $\begin{vmatrix} 1 & -\xi & \xi^2 \\ a_1 & h_1 & b_1 \\ a_2 & h_2 & b_2 \end{vmatrix} = 0.$

CHAPTER XIII

LINE-COORDINATES

89. Line-coordinates

Up to now we have regarded the point as fundamental but in this chapter we change our attitude and take the straight line as the fundamental geometric entity.

The straight line $lx + my + nz = 0$ determines and is determined by the two ratios $l:m:n$. Accordingly we may say that l, m and n are **homogeneous line-coordinates** of the straight line $lx + my + nz = 0$. We shall refer to it as the line (l, m, n). We also write the line $a \equiv (l_a, m_a, n_a)$.

The equation $x_A l + y_A m + z_A n = 0$ states that the point A lies on the line (l, m, n). Consequently this linear equation in l, m and n defines the pencil of straight lines with vertex at A. From our new viewpoint we may also say that this linear equation defines the point A. In line-coordinates a point is to be regarded as the vertex of the pencil formed by all straight lines through it.

Two linear equations,

$$x_A l + y_A m + z_A n = 0 \text{ and } x_B l + y_B m + z_B n = 0,$$

represent respectively the pencils of lines through A and B. The solution of these two equations must then represent the line common to both pencils. That is, the straight line AB. Hence these two equations represent the straight line $(y_A z_B - y_B z_A, z_A x_B - x_A z_B, x_A y_B - x_B y_A)$.

Three linear equations, $x_A l + y_A m + z_A n = 0, x_B l + y_B m + z_B n = 0$ and $x_C l + y_C m + z_C n = 0$, represent three pencils of lines through A, B and C respectively. These pencils have a straight line in common if the three points A, B, C are collinear and the necessary and sufficient condition (§3) is

$$\begin{vmatrix} x_A & y_A & z_A \\ x_B & y_B & z_B \\ x_C & y_C & z_C \end{vmatrix} = 0.$$

Example 1. Find the line-coordinates of the straight line joining the points: (*a*) (1, 2) and (– 2, 3), (*b*) (1, 0) and (0, 1), (*c*) (2, 0) and (– 1, 0).

Example 2. Calculate the point-coordinates of the point of intersection of the straight lines (*a*) (4, 1, – 1) and (1, – 1, – 4), (*b*) (3, – 2, 0) and (2, – 1, – 2).

90. Line-conic

We now ask: "What is the geometrical interpretation of a quadratic equation in the line-coordinates l, m and n?"

In §70 we proved that the straight line $lx + my + nz = 0$ is a tangent to the conic

$$S \equiv ax^2 + 2hxy + by^2 + 2gx + 2fy + c = 0, \tag{90.1}$$

if

$$\Sigma(l, m, n) \equiv \Sigma \equiv \mathscr{A}l^2 + \mathscr{B}m^2 + \mathscr{C}n^2 + 2\mathscr{F}mn + 2\mathscr{G}nl + 2\mathscr{H}lm = 0. \tag{90.2}$$

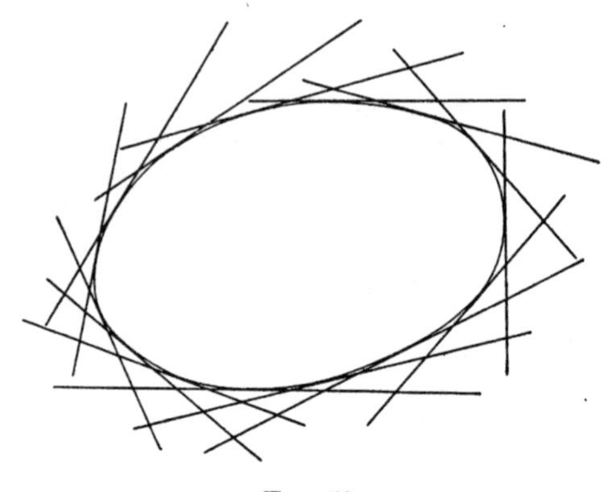

FIG. 50

Consequently a quadratic equation in l, m and n determines all the tangent lines of a conic (Fig. 50). We refer to the totality of tangent lines of a conic as a **line-conic.** Sometimes we say that (90.2) is the **tangential** equation of the conic. At other times we say that the **envelope** of the family of straight lines $lx + my + nz = 0$, subject to the condition $\Sigma = 0$ is the conic $S = 0$, and that $\Sigma = 0$ is the **line-equation** of the conic.

If a conic is given by its tangential equation (90.2), we may readily calculate† its point equation (90.1) by means of the relations (67.3).

† In returning to point coordinates, it is not necessary to calculate Δ. We have, in fact, if $\Delta \neq 0$ that the line-conic $\Sigma = 0$ yields in point coordinates the equation

$$(\mathscr{B}\mathscr{C} - \mathscr{F}^2)x^2 + 2(\mathscr{F}\mathscr{G} - \mathscr{H}\mathscr{C})xy + (\mathscr{A}\mathscr{C} - \mathscr{G}^2)y^2 + 2(\mathscr{H}\mathscr{F} - \mathscr{B}\mathscr{G})x + 2(\mathscr{H}\mathscr{G} - \mathscr{A}\mathscr{F})y + \mathscr{A}\mathscr{B} - \mathscr{H}^2 = 0.$$

An exceptional case rises when the determinant

$$\begin{vmatrix} \mathscr{A} & \mathscr{H} & \mathscr{G} \\ \mathscr{H} & \mathscr{B} & \mathscr{F} \\ \mathscr{G} & \mathscr{F} & \mathscr{C} \end{vmatrix} = 0.$$

Then Σ can be expressed as the product of two linear factors, each of which represents a point. Thus a line-conic degenerates to two points, or more particularly to the pencils of straight lines (Fig. 51) through these points. These two points, called a point-pair, may be real and distinct, coincident or complex conjugate.

When the line-conic is degenerate, it is not possible to proceed from (90.2) to (90.1) by means of (67.3).

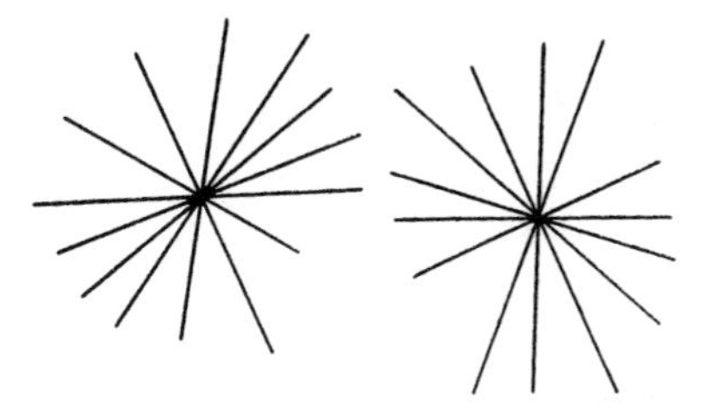

FIG. 51

Example 3. A straight line makes with the coordinate axes a triangle of constant area $2c^2$. Find the line-equation of its envelope and further obtain the corresponding point-equation.

Example 4. If the algebraic sum of the intercepts on the coordinate axes of a straight line is the constant c, obtain the line-equation and the point-equation of its envelope.

91. Joachimsthal's quadratic equation

We proceed to investigate the line-conic $\Sigma = 0$ (given by equation (90.2)) by the methods employed in §§68–70. We go into detail because the line-coordinate concept is novel at this stage.

The two straight lines $a \equiv (l_a, m_a, n_a)$ and $b \equiv (l_b, m_b, n_b)$ establish the pencil of lines $(\mu l_a + \lambda l_b,\ \mu m_a + \lambda m_b,\ \mu n_a + \lambda n_b)$. (Compare with (88.1), but there we have not made the multipliers homogeneous.) The lines of the pencil which touch the conic correspond to the roots of the equation

$$\begin{aligned}\mathscr{A}(\mu l_a + \lambda l_b)^2 + \mathscr{B}(\mu m_a + \lambda m_b)^2 + \mathscr{C}(\mu n_a + \lambda n_b)^2 \\ + 2\mathscr{F}(\mu m_a + \lambda m_b)(\mu n_a + \lambda n_b) + 2\mathscr{G}(\mu n_a + \lambda n_b)(\mu l_a + \lambda l_b) \\ + 2\mathscr{H}(\mu l_a + \lambda l_b)(\mu m_a + \lambda l_b) = 0,\end{aligned}$$

which simplifies to Joachimsthal's quadratic equation

$$\mu^2\Sigma_a + 2\mu\lambda\Phi_{ab} + \lambda^2\Sigma_b = 0, \tag{91.1}$$

where,

$$\Sigma_a = \Sigma(l_a, m_a, n_a); \ \Sigma_b = \Sigma(l_b, m_b, n_b);$$

$$\begin{aligned}\Phi_{ab} = \Phi_{ba} &\equiv (\mathscr{A}l_a + \mathscr{H}m_a + \mathscr{G}n_a)l_b + (\mathscr{H}l_a + \mathscr{B}m_a + \mathscr{F}n_a)m_b \\ &\quad + (\mathscr{G}l_a + \mathscr{F}m_a + \mathscr{C}n_a)n_b \\ &\equiv (\mathscr{A}l_b + \mathscr{H}m_b + \mathscr{G}n_b)l_a + (\mathscr{H}l_b + \mathscr{B}m_b + \mathscr{F}n_b)m_a \\ &\quad + (\mathscr{G}l_b + \mathscr{F}m_b + \mathscr{C}n_b)n_a.\end{aligned}$$

Equation (91.1) has two roots in the ratio λ/μ and these correspond to the two lines c and d (Fig. 52) of the pencil which touch the conic.

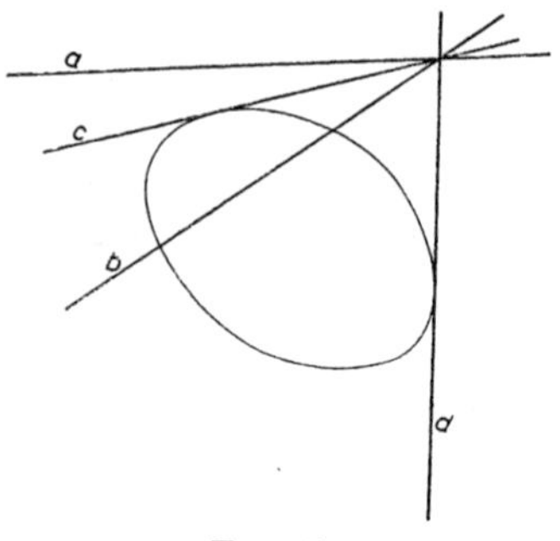

FIG. 52

Example 5. Obtain the line-coordinates of the pair of tangent lines to the conic $l^2 + 4m^2 - 2n^2 + 14lm + 2mn + 8nl = 0$ through the point of intersection of the lines $(1, 1, -2)$ and $(8, -1, 2)$.

92. Polar properties

As the straight line is the fundamental entity in line-coordinates, we begin by defining two lines a and b (Fig. 52) to be conjugate with respect to the line-conic if the two tangent lines c and d of their pencil divide a and b harmonically. From §88, we see that the sum of the roots of (91.1) is zero. Hence, the necessary and sufficient condition that the two lines a and b be conjugate is $\Phi_{ab} = 0$.

We next define the pole of a line a with respect to a line-conic $\Sigma = 0$ to be the assemblage of all straight lines conjugate to a. Consequently the pole of the straight line is given by

$$\begin{aligned}\Phi_a \equiv (\mathscr{A}l_a + \mathscr{H}m_a + \mathscr{G}n_a)l + (\mathscr{H}l_a + \mathscr{B}m_a + \mathscr{F}n_a)m \\ + (\mathscr{G}l_a + \mathscr{F}m_a + \mathscr{C}n_a)n = 0. \quad (92.1)\end{aligned}$$

This equation is linear in l, m and n and so represents a point.

Conversely, the **polar** of a point is defined to be that line whose pole is the given point. Let us obtain the line-coordinates of the polar a of the point $x_A l + y_A m + z_A n = 0$. The pole of the line a is given by (92.1) which must then represent the same point as $x_A l + y_A m + z_A n = 0$. The comparison of coefficients yields

$$\mathscr{A} l_a + \mathscr{H} m_a + \mathscr{G} n_a = \kappa x_A,$$
$$\mathscr{H} l_a + \mathscr{B} m_a + \mathscr{F} n_a = \kappa y_A,$$
$$\mathscr{G} l_a + \mathscr{F} m_a + \mathscr{C} n_a = \kappa z_A,$$

where κ is some factor of proportionality. Multiply these equations in turn by a, h, g; then by h, b, f; and again by g, f, c and add. In virtue of (67.2) the results are

$$\Delta l_a = \kappa(a x_A + h y_A + g z_A),$$
$$\Delta m_a = \kappa(h x_A + b y_A + f z_A),$$
$$\Delta n_a = \kappa(g x_A + f y_A + c z_A).$$

Thus the polar of the point $x_A l + y_A m + z_A n = 0$ is the straight line

$$(a x_A + h y_A + g z_A,\ h x_A + b y_A + f z_A,\ g x_A + f y_A + c z_A). \quad (92.2)$$

Consider the two points

$$x_A l + y_A m + z_A n = 0 \text{ and } x_B l + y_B m + z_B n = 0.$$

The reader is asked to show that each point lies on the polar of the other if $T_{AB} = 0$ (see §68 for the definition of T_{AB}). Two points, such that each lies on the polar of the other are called **conjugate** points.

We note that this section repeats the results of §69. The difference lies merely in the point of view.

Example 6. Obtain the line-coordinates of the polar of the point $l + m - n = 0$ with respect to the conic of Example 5.

93. Intersections with conic

Let a be a tangent line of the line-conic $\Sigma = 0$ (Fig. 53). That is $\Sigma_a = 0$. Then Joachimsthal's equation (91.1) reduces to $2\mu\lambda\Phi_{ab} + \lambda^2\Sigma_b = 0$. Choose the point of contact of the line a as the vertex of a pencil of lines. The lines of this pencil which touch the conic are coincident and so $\Phi_{ab} = 0$, if the line b passes through the point of contact of the tangent line a. That is, the point of contact of the tangent line a has the equation $\Phi_a = 0$. Comparing with (92.1) we see once more that the pole of a tangent line is at its point of contact.

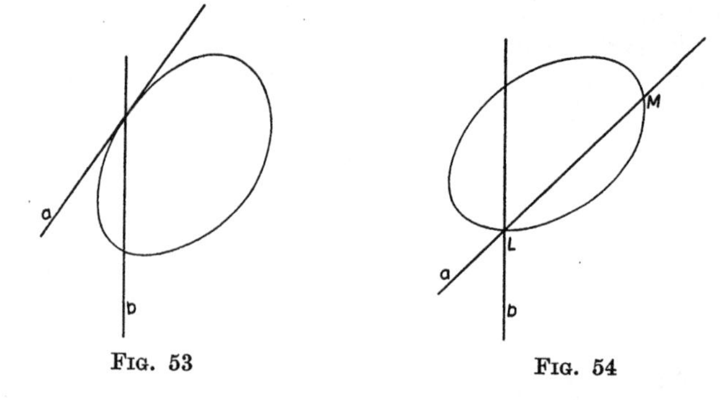

FIG. 53 FIG. 54

If the lines a and b intersect on the conic (Fig. 54) but are not tangent lines, then Joachimsthal's equation (91.1) has equal roots. That is, $\Sigma_a\Sigma_b - \Phi_{ab}^2 = 0$. The totality of all lines b which intersect both a and the conic form two pencils at the points L and M. Hence the two points of intersection of a line a and a line-conic $\Sigma = 0$ are given by

$$\Sigma_a\Sigma - \Phi_a^2 = 0.$$

Example 7. Obtain the equations of the points of intersection of the straight line $(1, 2, -1)$ and the conic $4l^2 - 9m^2 - 29n^2 - 12lm - 30mn - 16nl = 0$.

Example 8. Verify that the line $(2, -5, 1)$ is a tangent line of the conic in Example 7 and calculate the equation of the point of contact.

94. Principle of duality

We have considered the equation $lx + my + nz = 0$ from two points of view. In point-coordinates, we take l, m, n as constants and $lx + my + nz = 0$ represents all the points (x, y, z) which lie on the given line (l, m, n). On the other hand, in line-coordinates we take x, y, z as constants and $lx + my + nz = 0$ represents all the lines which pass through the fixed point (x, y, z). Thus there is a duality between a straight line considered as a range of points and a point considered as the pencil of lines through the point as vertex.

An algebraical result then has two interpretations according as we choose x, y, z as point coordinates or as line-coordinates. Two such results are said to be dual to one another, and the principle which allows this is called the principle of duality.

As an illustration we repeat §78 but use l, m, n instead of x, y, z to emphasize the fact that we now interpret from the point of view of line-coordinates. The reader is asked to compare the following section with §78 very carefully.

95. Range of conics

Consider the two line-conics

$$\Sigma \equiv \mathscr{A}l^2 + \mathscr{B}m^2 + \mathscr{C}n^2 + 2\mathscr{F}mn + 2\mathscr{G}nl + 2\mathscr{H}lm = 0,$$
$$\Sigma' \equiv \mathscr{A}'l^2 + \mathscr{B}'m^2 + \mathscr{C}'n^2 + 2\mathscr{F}'mn + 2\mathscr{G}'nl + 2\mathscr{H}'lm = 0,$$

and set up the equation

$$\Sigma + \lambda\Sigma' = 0. \tag{95.1}$$

This equation is of the second degree in l, m and n and thus represents a line-conic, which must touch the common tangents of the line-conics Σ and Σ' given by $\Sigma = \Sigma' = 0$. These two equations are of the second degree and so in general Σ and Σ' have four common tangents. Consequently $\Sigma + \lambda\Sigma' = 0$ represents a system of conics touching four given straight lines called **base lines.** Such a system is called a **range** of conics.

Touching any line, not a base line, there is one and only one conic of the range because (95.1) yields a linear equation in λ.

Two conics of a range pass through a given point because this condition, obtained by substitution in the point-equation corresponding to (95.1), is quadratic in λ.

The equation of a range of conics touching the four given straight lines a, b, c and d is

$$U_1U_2 + \lambda U_3U_4 = 0,$$

where $U_1 = 0$, $U_2 = 0$, $U_3 = 0$, $U_4 = 0$ are the equations of the points of intersection of a and b, c and d, a and d, b and c respectively.

The line-conic $\Sigma + \lambda\Sigma' = 0$ degenerates to a point-pair if

$$\begin{vmatrix} \mathscr{A} + \lambda\mathscr{A}' & \mathscr{H} + \lambda\mathscr{H}' & \mathscr{G} + \lambda\mathscr{G}' \\ \mathscr{H} + \lambda\mathscr{H}' & \mathscr{B} + \lambda\mathscr{B}' & \mathscr{F} + \lambda\mathscr{F}' \\ \mathscr{G} + \lambda\mathscr{G}' & \mathscr{F} + \lambda\mathscr{F}' & \mathscr{C} + \lambda\mathscr{C}' \end{vmatrix} = 0.$$

This is a cubic equation in λ. Accordingly there are three point-pairs in a range of conics. In Fig. 55 the degenerate point-pairs are M, P; N, L and Q, R.

Example 9. How many (a) rectangular hyperbolas, (b) parabolas are there in a range of conics?

Example 10. Show that the centre of the conic $\Sigma + \lambda\Sigma' = 0$ is at the point $(\mathscr{G} + \lambda\mathscr{G}')l + (\mathscr{F} + \lambda\mathscr{F}')m + (\mathscr{C} + \lambda\mathscr{C}')n = 0$. Deduce that the locus of the centres of conics which are inscribed in a quadrilateral is the straight line joining the mid-points of its diagonals.

96. Foci

By the use of line-coordinates it is possible to obtain the foci of a conic in the following elegant manner. We proved in §76 that the foci are the points of intersection of the tangents from the circular points I and J to the conic. We depict the situation symbolically in Fig. 56. This figure is not factual as we cannot draw the line at infinity nor even complex points. We see then that all conics inscribed in the quadrilateral $FGF'G'$ have the same foci. Such conics are said to be confocal and they form a range. One degenerate line-conic of this range is the point-pair I and J, whose line equation is $(l + im)(l - im) \equiv l^2 + m^2 = 0$. Hence the range of conics confocal with $S = 0$ is given in line-coordinates by

$$\Sigma + \lambda(l^2 + m^2) = 0. \tag{96.1}$$

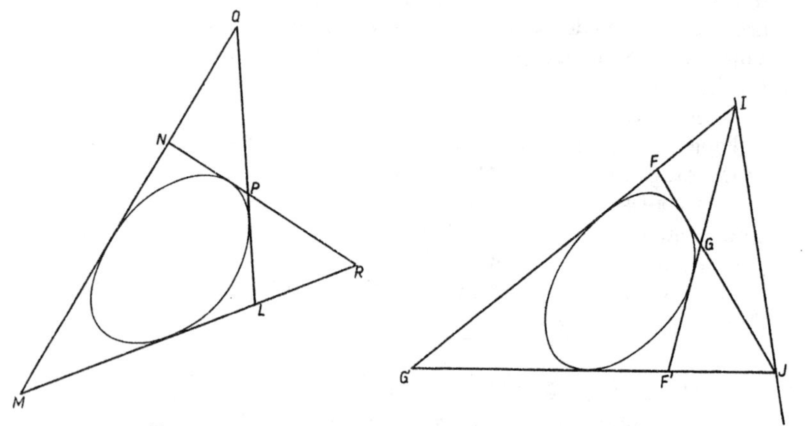

FIG. 55 FIG. 56

This line-conic is degenerate if the determinant

$$\begin{vmatrix} \mathscr{A} + \lambda & \mathscr{H} & \mathscr{G} \\ \mathscr{H} & \mathscr{B} + \lambda & \mathscr{F} \\ \mathscr{G} & \mathscr{F} & \mathscr{C} \end{vmatrix} = 0,$$

which reduces to

$$\mathscr{C}\lambda^2 + (\mathscr{C}\mathscr{A} - \mathscr{G}^2 + \mathscr{B}\mathscr{C} - \mathscr{F}^2)\lambda + \begin{vmatrix} \mathscr{A} & \mathscr{H} & \mathscr{G} \\ \mathscr{H} & \mathscr{B} & \mathscr{F} \\ \mathscr{G} & \mathscr{F} & \mathscr{C} \end{vmatrix} = 0.$$

In virtue of (67.3) we obtain the quadratic equation

$$J\lambda^2 + I\Delta\lambda + \Delta^2 = 0, \tag{96.2}$$

where I and J are now the relative invariants of S. The discriminant is $I^2\Delta^2 - 4\Delta^2 J = \Delta^2[(a - b)^2 + 4h^2] > 0$ and so both roots λ_1 and λ_2 (say) are real.

The degenerate line-conic $\Sigma + \lambda_1(l^2 + m^2) = 0$ has real† factors if $\mathscr{G}^2 - (\mathscr{A} + \lambda_1)\mathscr{C} > 0$. By (67.3) this condition is $b\Delta + \lambda_1 J < 0$. We have

$$\begin{aligned}(b\Delta + \lambda_1 J)(b\Delta + \lambda_2 J) &= b^2\Delta^2 + b\Delta J(\lambda_1 + \lambda_2) + J^2\lambda_1\lambda_2,\\ &= b^2\Delta^2 + b\Delta J(-I\Delta)/J + J^2\Delta^2/J,\\ &= -h^2\Delta^2,\\ &< 0.\end{aligned}$$

Accordingly one factor (say) $b\Delta + \lambda_1 J > 0$, whilst the other $b\Delta + \lambda_2 J < 0$. Thus factors of $\Sigma + \lambda_1(l^2 + m^2)$ are real and the factors of $\Sigma + \lambda_2(l^2 + m^2)$ are complex conjugate.

To complete our task we factorize

$$\Sigma + \lambda_1(l^2 + m^2) \equiv (\alpha_1 l + \beta_1 m + \gamma_1 n)(\alpha_2 l + \beta_2 m + \gamma_2 n)$$

and we see that the real foci are at the points $(\alpha_1, \beta_1, \gamma_1)$ and $(\alpha_2, \beta_2, \gamma_2)$.

If the conic is a parabola, $J = 0$ and the quadratic (96.2) reduces to a linear equation with the one root $\lambda = -\Delta/I$. The foci are the pair of points of the degenerate line-conic

$$\mathscr{A}l^2 + \mathscr{B}m^2 + 2\mathscr{F}mn + 2\mathscr{G}nl + 2\mathscr{H}lm - \frac{\Delta}{I}(l^2 + m^2) = 0.$$

There is no term in n^2 and so the factors are of the type

$$(\alpha_1 l + \beta_1 m + \gamma_1 n)(\alpha_2 l + \beta_2 m).$$

The real finite focus is at $(\alpha_1, \beta_1, \gamma_1)$ and there is another real focus at the point at infinity $(\alpha_2, \beta_2, 0)$. This focus at infinity of the parabola was not apparent from our previous work in §76.

Example 11. Find the foci of the conics in Example 16, Chapter X, by the method of this section.

Example 12. Show that the focus at infinity of a parabola is the point at infinity on its axis.

97. Confocal conics

In the last section we saw that all the conics confocal to $S = 0$ form the range of line-conics

$$\Sigma + \lambda(l^2 + m^2) = 0.$$

---

† The degenerate line-conic $\Sigma = 0$ is real, if the tangents from the point $m = 0$ are real. That is $\mathscr{A}l^2 + \mathscr{C}n^2 + 2\mathscr{G}nl = 0$ has real roots and so $\mathscr{G}^2 - \mathscr{A}\mathscr{C} > 0$.

We return to point coordinates (see footnote to §90), and the equation is

$$(\mathscr{B}\mathscr{C} - \mathscr{F}^2 + \mathscr{C}\lambda)x^2 + (C\mathscr{A} - \mathscr{G}^2 + \mathscr{C}\lambda)y^2 + 2(\mathscr{F}\mathscr{G} - \mathscr{H}\mathscr{C})xy$$
$$+ 2(\mathscr{H}\mathscr{F} - \mathscr{B}\mathscr{G} - \lambda\mathscr{G})x + 2(\mathscr{H}\mathscr{G} - \mathscr{A}\mathscr{F} - \lambda\mathscr{F})y$$
$$+ (\mathscr{A}\mathscr{B} - \mathscr{H}^2 + (\mathscr{A} + \mathscr{B})\lambda + \lambda^2) = 0,$$

which reduces in virtue of (67.3) to

$$\Delta S + \lambda\Theta + \lambda^2 = 0, \tag{97.1}$$

where Θ is defined by (71.1). ($\Theta = 0$ is the orthoptic circle of $S = 0$.)

In tackling examples the reader is advised to work without quoting this formula as in §§98 and 99.

Example 13. Obtain in point-coordinates the range of conics confocal to the following conics: (*a*) $2xy = c^2$, (*b*) $x^2 + y^2 + (x \cos \theta + y \sin \theta)^2 = 1$.

98. Central confocal conics

It is convenient to choose the equation of the central conic as

$$\frac{x^2}{\alpha} + \frac{y^2}{\beta} - 1 = 0$$

and the corresponding tangential equation is

$$\alpha l^2 + \beta m^2 - 1 = 0.$$

The confocal conics are the members of the range

$$(\alpha + \lambda)l^2 + (\beta + \lambda)m^2 - 1 = 0, \tag{98.1}$$

and their corresponding point equation is

$$\frac{x^2}{\alpha + \lambda} + \frac{y^2}{\beta + \lambda} - 1 = 0. \tag{98.2}$$

Let us suppose that $\alpha > \beta > 0$. Then the confocal is an ellipse for $\lambda > -\beta$, a hyperbola for $-\beta > \lambda > -\alpha$ and a virtual ellipse for $\lambda < -\alpha$. If $\lambda = -\alpha$ or $-\beta$ we cannot proceed from (98.1) to (98.2) as the line-conic degenerates to a pair of foci.

To find the confocals through a given point P, we have the quadratic in λ,

$$f(\lambda) \equiv (\beta + \lambda)x_P{}^2 + (\alpha + \lambda)y_P{}^2 - (\alpha + \lambda)(\beta + \lambda) = 0.$$

For $\lambda = -\alpha, -\beta$ and $+\infty, f(\lambda)$ is $-$ve, $+$ve and $-$ve respectively and so the two roots λ_1, λ_2 are such that $-\beta > \lambda_1 > -\alpha$ and

$\lambda_2 > -\beta$. Thus there are two confocals through a real point, one an ellipse and the other a hyperbola.

The tangents at P to the two confocals are

$$\frac{x_P x}{\alpha + \lambda_1} + \frac{y_P y}{\beta + \lambda_1} - 1 = 0; \quad \frac{x_P x}{\alpha + \lambda_2} + \frac{y_P y}{\beta + \lambda_2} - 1 = 0.$$

But

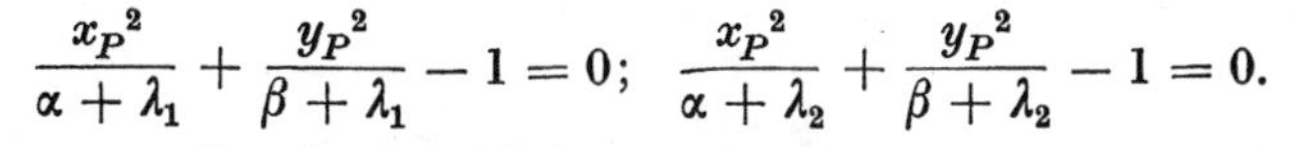

$$\frac{x_P^2}{\alpha + \lambda_1} + \frac{y_P^2}{\beta + \lambda_1} - 1 = 0; \quad \frac{x_P^2}{\alpha + \lambda_2} + \frac{y_P^2}{\beta + \lambda_2} - 1 = 0.$$

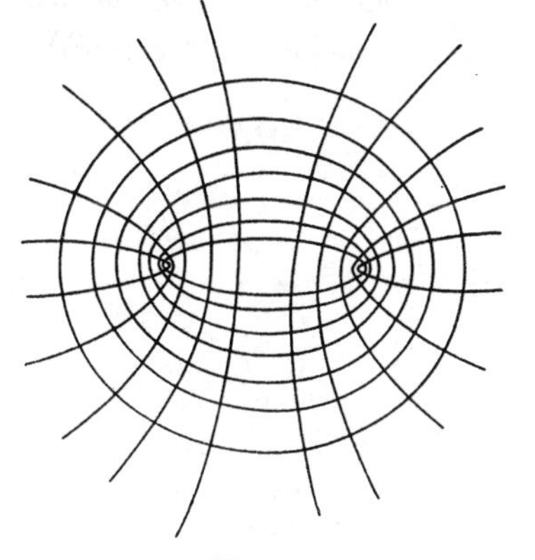

FIG. 57

By subtraction, and subsequent division by $\lambda_1 - \lambda_2 \neq 0$, we have

$$\frac{x_P^2}{(\alpha + \lambda_1)(\alpha + \lambda_2)} + \frac{y_P^2}{(\beta + \lambda_1)(\beta + \lambda_2)} = 0.$$

That is, the tangents to the confocals are orthogonal. Therefore, the two confocals through a point cut each other orthogonally.

A family of confocal conics is depicted in Fig. 57.

Example 14. Prove that the locus of the poles of a given straight line with respect to the conics of a central confocal system of conics is the normal at the point of contact to the conic of the system which touches the given straight line.

Example 15. Show that the locus of the point of intersection of two perpendicular tangents, one to each of two given confocal central conics is a circle.

99. Confocal parabolas

We now seek the conics confocal to the parabola $y^2 = 4ax$, whose tangential equation is $2am^2 - 2ln = 0$. The confocals are the members of the range

$$\lambda l^2 + (2a + \lambda)m^2 - 2ln = 0,$$

and the corresponding point equation is

$$y^2 = (2a + \lambda)(2x + \lambda).$$

Make the substitution $2a + \lambda = 2\mu$ and translate to parallel axes through the point $(a, 0)$, that is the common focus, as origin. Then we obtain the more usual equation of the confocals, namely

$$y^2 = 4\mu(x + \mu). \tag{99.1}$$

The confocal parabolas (Fig. 58) form two systems according as $\mu > 0$ when the vertex is to the left, or $\mu < 0$ when the vertex is to the right.

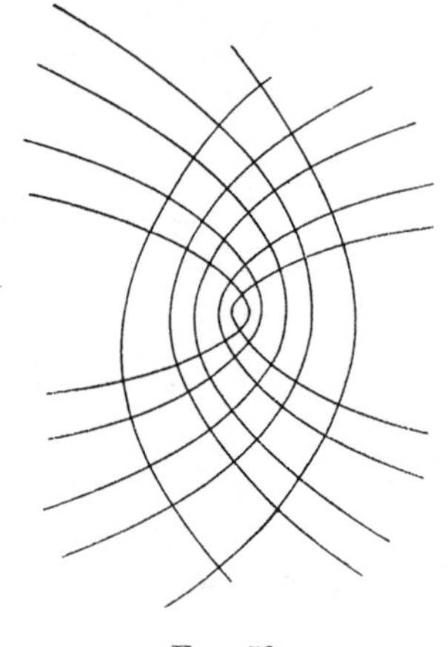

Fig. 58

To find the confocals through a given point P, we have the quadratic equation $4\mu^2 + 4\mu x_P - y_P{}^2 = 0$. The product of the roots is negative and so two confocals, one of each system, pass through a point. Let the roots of this quadratic equation be μ_1 and μ_2. Then

$$4\mu_1{}^2 + 4\mu_1 x_P - y_P{}^2 = 0; \quad 4\mu_2{}^2 + 4\mu_2 x_P - y_P{}^2 = 0.$$

Therefore

$$\mu_2(4\mu_1{}^2 + 4\mu_1 x_P - y_P{}^2) - \mu_1(4\mu_2{}^2 + 4\mu_2 x_P - y_P{}^2) = 0,$$

which reduces on division by $\mu_1 - \mu_2 \neq 0$ to $4\mu_1\mu_2 + y_P{}^2 = 0$. Thus the two tangents at P,

$$yy_P = 2\mu_1(x + x_P) + 4\mu_1{}^2; \quad yy_P = 2\mu_2(x + x_P) + 4\mu_2{}^2$$

are mutually orthogonal.

Example 16. Show that the result of Example 14 is also valid for a system of confocal parabolas.

Example 17. For a system of confocal parabolas, show that the locus in Example 15 becomes a straight line perpendicular to the common axis.

Miscellaneous examples

1. Obtain the necessary and sufficient conditions that the two conics $ax^2 + 2hxy + by^2 = 1$, $\lambda x^2 + 2\mu xy + \nu y^2 = 1$ be confocal.

Further, find the condition that the latter conic can be placed so as to be a member of the confocal family determined by the former conic.

2. Find the necessary and sufficient condition that the equation $\mathscr{A}l^2 + \mathscr{B}m^2 + \mathscr{C}n^2 + 2\mathscr{F}mn + 2\mathscr{G}nl + 2\mathscr{H}lm = 0$ be a circle. Further, show that the equation of its centre is $l/\mathscr{F} + m/\mathscr{G} + n/\mathscr{H} = 0$.

3. Prove that the conics $(\alpha y - \beta x)^2 = p(y^2 - 2\beta y)$ and $(\alpha y - \beta x)^2 = q(x^2 - 2\alpha x)$ are confocal if $p + q = 0$.

4. Show that the equation $pmn + qnl + rlm = 0$ represents a parabola and find the equation of its focus.

5. Obtain the equation of the directrix of the parabola which touches the four common tangents of the conics $ax^2 + by^2 + 2gx + c = 0$ and $ax^2 + by^2 + 2fy + c = 0$.

6. The line-equation of a conic is $\mathscr{A}l^2 + \mathscr{B}m^2 + \mathscr{C}n^2 + 2\mathscr{F}mn + 2\mathscr{G}nl + 2\mathscr{H}lm = 0$. Interpret the following equations:

(a) $\mathscr{A}l^2 + 2\mathscr{H}lm + \mathscr{B}m^2 = 0$,

(b) $\mathscr{G}l + \mathscr{F}m + \mathscr{C}n = 0$,

(c) $(\mathscr{A}\mathscr{C} - \mathscr{G}^2)l^2 - 2(\mathscr{F}\mathscr{G} - \mathscr{H}\mathscr{C})lm + (\mathscr{B}\mathscr{C} - \mathscr{F}^2)m^2 = 0$.

7. Prove that the orthoptic circles of all conics of a range form a coaxal system and that the directrix of the parabola of the range is the radical axis.

8. Show that the envelope of the polar of a fixed point with respect to the conics of a confocal system is a parabola.

Answers

1. (a) $(1, 3, -7)$, (b) $(1, 1, -1)$, (c) $(0, 1, 0)$. **2.** (a) $(1, -3)$, (b) $(4, 6)$. **3.** $4c^2lm - n^2 = 0$, $xy = c^2$. **4.** $clm + mn + nl = 0$, $(x - y)^2 - 2c(x + y) + c^2 = 0$. **5.** $(4, 1, -2)$, $(2, -1, 2)$. **6.** $(2, -3, 7)$. **7.** $l + n = 0$, $l - 3m - 5n = 0$. **8.** $5l + 3m + 5n = 0$. **9.** (a) two, (b) one. **13.** (a) $2c^2xy - \lambda(x^2 + y^2) + \lambda^2 - c^4 = 0$, (b) $2(\lambda - 1)(x^2 + y^2) - 2(x \cos \theta + y \sin \theta)^2 + \lambda^2 - 3\lambda + 2 = 0$.

Miscellaneous examples. **1.** $\dfrac{a - b}{\lambda - \nu} = \dfrac{ab - h^2}{\lambda\nu - \mu^2} = \dfrac{h}{\mu}$; $(ab - h^2)^2[(\lambda + \mu)^2 - 4(\lambda\nu - \mu^2)] = (\lambda\nu - \mu^2)^2[(a + b)^2 - 4(ab - h^2)]$. **2.** $\mathscr{B}\mathscr{C} - \mathscr{F}^2 = \mathscr{C}\mathscr{A} - \mathscr{G}^2$ and $\mathscr{F}\mathscr{G} - \mathscr{H}\mathscr{C} = 0$. **4.** $rpl + qrm + (p^2 + q^2)n = 0$. **5.** $2(bgx - afy) + f^2 - g^2 = 0$. **6.** (a) The points at infinity on the tangents through the origin, (b) centre, (c) the two points at infinity on the conic.

CHAPTER XIV

GENERALIZED HOMOGENEOUS COORDINATES

In this chapter we shall manipulate most of the algebraic work in point coordinates. We will, however, indicate the dual results (see §94) obtained by interpreting the coordinates as line-coordinates.

100. Cofactor notation

In contrast to §67, we introduce the unsymmetrical determinant

$$R \equiv \begin{vmatrix} a_1 & b_1 & c_1 \\ a_2 & b_2 & c_2 \\ a_3 & b_3 & c_3 \end{vmatrix} \neq 0. \tag{100.1}$$

Further, we define the cofactors

$$\left.\begin{aligned} A_1 &\equiv b_2c_3 - b_3c_2; \quad B_1 \equiv c_2a_3 - c_3a_2; \quad C_1 \equiv a_2b_3 - a_3b_2; \\ A_2 &\equiv b_3c_1 - b_1c_3; \quad B_2 \equiv c_3a_1 - c_1a_3; \quad C_2 \equiv a_3b_1 - a_1b_3; \\ A_3 &\equiv b_1c_2 - b_2c_1; \quad B_3 \equiv c_1a_2 - c_2a_1; \quad C_3 \equiv a_1b_2 - a_2b_1. \end{aligned}\right\} \tag{100.2}$$

The reader is asked to verify the identities

$$\left.\begin{aligned} a_1A_1 + a_2A_2 + a_3A_3 = R; \quad a_1B_1 + a_2B_2 + a_3B_3 &= 0; \\ a_1C_1 + a_2C_2 + a_3C_3 &= 0; \\ b_1A_1 + b_2A_2 + b_3A_3 = 0; \quad b_1B_1 + b_2B_2 + b_3B_3 &= R; \\ b_1C_1 + b_2C_2 + b_3C_3 &= 0; \\ c_1A_1 + c_2A_2 + c_3A_3 = 0; \quad c_1B_1 + c_2B_2 + c_3B_3 &= 0; \\ c_1C_1 + c_2C_2 + c_3C_3 &= R. \end{aligned}\right\} \tag{100.3}$$

101. Generalized homogeneous coordinates

Consider the three linear equations

$$\left.\begin{aligned} \theta\xi &= a_1x + b_1y + c_1z, \\ \theta\eta &= a_2x + b_2y + c_2z, \\ \theta\zeta &= a_3x + b_3y + c_3z, \end{aligned}\right\} \tag{101.1}$$

where θ is a factor of proportionality, and R defined by (100.1) is not zero. Multiply these three equations in turn by A_1, A_2, A_3, then by B_1, B_2, B_3 and finally by C_1, C_2, C_3. In virtue of (100.3) the additions yield

$$\left.\begin{aligned} \varphi x &= A_1\xi + A_2\eta + A_3\zeta, \\ \varphi y &= B_1\xi + B_2\eta + B_3\zeta, \\ \varphi z &= C_1\xi + C_2\eta + C_3\zeta, \end{aligned}\right\} \tag{101.2}$$

where $\varphi = R/\theta$.

We see that the ratios $x:y:z$ determine and are uniquely determined by the ratios $\xi:\eta:\zeta$. We can, therefore, regard ξ, η, ζ as point coordinates, called **generalized homogeneous coordinates.**

The triple (0, 0, 0) is excluded from cartesian homogeneous coordinates and so it is also excluded from generalized homogeneous coordinates.

The line at infinity $z = 0$ now takes the more general equation $C_1\xi + C_2\eta + C_3\zeta = 0$. It appears that the line at infinity is going to lose its special significance.

Dually, three linear equations (101.1) with x, y, z replaced by l, m, n and ξ, η, ζ by λ, μ, ν yield the generalized homogeneous line-coordinates λ, μ, ν. The coordinates of the line at infinity are now (c_1, c_2, c_3).

102. Geometrical significance

The three straight lines $\xi = 0$, $\eta = 0$ and $\zeta = 0$, where ξ, η and ζ are defined by (101.1), form a triangle because the determinant R is not zero by hypothesis. In Fig. 59, let p, q and r be the perpendicular distances on these lines from the point (x, y, z). Since we can always select $z = 1$ for finite points, we have

$$p = (a_1x + b_1y + c_1z)/\sqrt{a_1^2 + b_1^2}.$$

Consequently

$$\theta\xi = k_1p; \quad \theta\eta = k_2q; \quad \theta\zeta = k_3r,$$

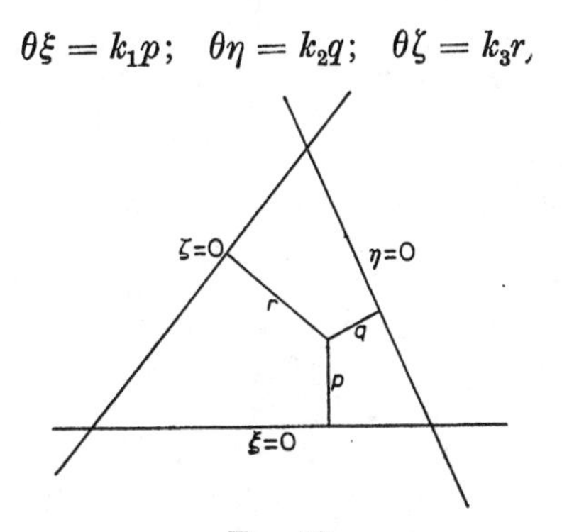

FIG. 59

where

$$k_1^2 = a_1^2 + b_1^2; \quad k_2^2 = a_2^2 + b_2^2; \quad k_3^2 = a_3^2 + b_3^2.$$

Hence

$$\xi:\eta:\zeta = k_1 p:k_2 q:k_3 r.$$

That is, generalized homogeneous coordinates are three numbers ξ, η, ζ not all zero which are proportional to assigned multiples of the perpendiculars from the point to the sides of a given triangle, called the **triangle of reference.**

We do not try to dualize this result. However, the equations

$$a_1 l + b_1 m + c_1 n = 0, \quad a_2 l + b_2 m + c_2 n = 0, \quad a_3 l + b_3 m + c_3 n = 0$$

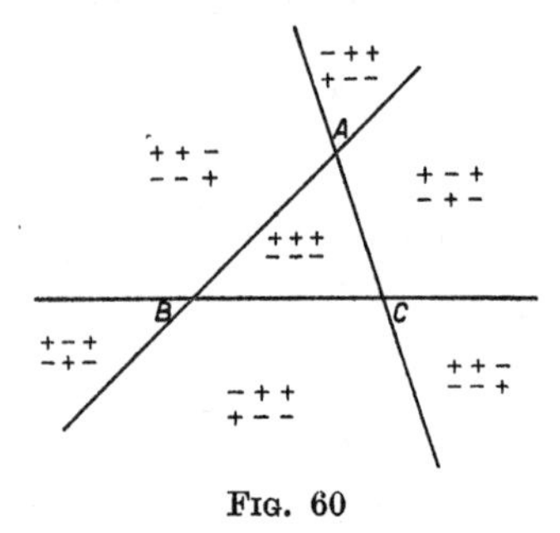

Fig. 60

determine three points which form the vertices of a triangle, called the triangle of reference with respect to generalized homogeneous line-coordinates.

From (101.1) we see that a system of generalized homogeneous coordinates involves eight effective constants, namely the ratios of the nine quantities a_1, b_1, c_1, a_2, b_2, c_2, a_3, b_3 and c_3. Fixing the side of a triangle of reference is equivalent to the determination of two ratios. Hence the selection of the triangle of reference yields six conditions. The remaining two conditions depend on the choice of the ratios $k_1:k_2:k_3$. In effect, this means that the two conditions depend on the choice of the point (1, 1, 1) called the **unit point.**

Dually; the selection of the vertices of the triangle of reference in line-coordinates gives six conditions and the remaining two conditions depend on the choice of the line (1, 1, 1) called the **unit line.**

Let us examine the effect of choosing the unit point U inside the triangle of reference ABC (Fig. 60). The signs of the triples ξ, η, ζ must then be as depicted in the figure. The reader is advised to draw diagrams of a similar nature when U is situated outside the triangle of reference.

Further, we have $A \equiv (1, 0, 0)$, $B \equiv (0, 1, 0)$, $C \equiv (0, 0, 1)$ and the equations of the sides BC, CA and AB of the triangle of reference are $\xi = 0$, $\eta = 0$ and $\zeta = 0$ respectively.

Example 1. Choose the unit-point at the centroid of the triangle of reference. The coordinates are then called **areal coordinates.** If a, b, c are the lengths of the sides opposite to the corresponding angles A, B, C of the triangle of reference, prove the following results:

(*a*) the in-centre of the triangle is at (a, b, c),

(*b*) the circumcentre of the triangle is at $(\sin 2A, \sin 2B, \sin 2C)$,

(*c*) the orthocentre of the triangle is at $(\tan A, \tan B, \tan C)$,

(*d*) the equation of the line at infinity is $x + y + z = 0$.

Example 2. Choose the in-centre of the triangle of reference as the unit-point. The coordinates are then called **trilinear coordinates.** Prove the following results:

(*a*) the centroid of the triangle is at $(1/a, 1/b, 1/c)$,

(*b*) the circumcentre of the triangle is at $(\cos A, \cos B, \cos C)$,

(*c*) the orthocentre of the triangle is at $(\sec A, \sec B, \sec C)$,

(*d*) the equation of the line at infinity is $ax + by + cz = 0$.

103. Range of points

Select two points $P \equiv (\xi_P, \eta_P, \zeta_P)$ and $Q \equiv (\xi_Q, \eta_Q, \zeta_Q)$ as the base points of a range of points. In virtue of (101.1) we have

$$\begin{aligned} \theta_P\xi_P &= a_1x_P + b_1y_P + c_1z_P, & \theta_Q\xi_Q &= a_1x_Q + b_1y_Q + c_1z_Q, \\ \theta_P\eta_P &= a_2x_P + b_2y_P + c_2z_P, & \theta_Q\eta_Q &= a_2x_Q + b_2y_Q + c_2z_Q, \\ \theta_P\zeta_P &= a_3x_P + b_3y_P + c_3z_P, & \theta_Q\zeta_Q &= a_3x_Q + b_3y_Q + c_3z_Q. \end{aligned}$$

Here we note that the multipliers θ may change from point to point.

Any point on the range PQ has cartesian coordinates $(\mu x_P + \lambda x_Q, \mu y_P + \lambda y_Q, \mu z_P + \lambda z_Q)$ and the corresponding generalized homogeneous coordinates are

$$\begin{aligned} \theta\xi &= a_1(\mu x_P + \lambda x_Q) + b_1(\mu y_P + \lambda y_Q) + c_1(\mu z_P + \lambda z_Q) \\ &= \mu\theta_P\xi_P + \lambda\theta_Q\xi_Q, \\ \theta\eta &= a_2(\mu x_P + \lambda x_Q) + b_2(\mu y_P + \lambda y_Q) + c_2(\mu z_P + \lambda z_Q) \\ &= \mu\theta_P\eta_P + \lambda\theta_Q\eta_Q, \\ \theta\zeta &= a_3(\mu x_P + \lambda x_Q) + b_3(\mu y_P + \lambda y_Q) + c_3(\mu z_P + \lambda z_Q) \\ &= \mu\theta_P\zeta_P + \lambda\theta_Q\zeta_Q. \end{aligned}$$

That is, the range of points has the coordinates

$$(\rho\xi_P + \tau\xi_Q, \rho\eta_P + \tau\eta_Q, \rho\zeta_P + \tau\zeta_Q) \tag{103.1}$$

if we set $\tau/\rho = \lambda\theta_Q/\mu\theta_P$. There is a homographic correspondence between τ/ρ and λ/μ, which preserves cross-ratio. Consequently the cross-ratio of the four collinear points $(\rho_i\xi_P + \tau_i\xi_Q, \rho_i\eta_P + \tau_i\eta_Q, \rho_i\zeta_P + \tau_i\zeta_Q)$ for $i = 1, 2, 3, 4$ is the cross-ratio of the four values of τ_i/ρ_i.

In particular, (ξ_P, η_P, ζ_P), (ξ_Q, η_Q, ζ_Q) and $(\rho\xi_P \pm \tau\xi_Q, \rho\eta_P \pm \tau\eta_Q, \rho\zeta_P \pm \tau\zeta_Q)$ form a harmonic range.

It follows easily from (103.1) that the three points (ξ_1, η_1, ζ_1), (ξ_2, η_2, ζ_2) and (ξ_3, η_3, ζ_3) are collinear if the determinant

$$\begin{vmatrix} \xi_1 & \eta_1 & \zeta_1 \\ \xi_2 & \eta_2 & \zeta_2 \\ \xi_3 & \eta_3 & \zeta_3 \end{vmatrix} = 0.$$

104. Straight line

In cartesian homogeneous coordinates, the straight line is represented by $lx + my + nz = 0$. Using (101.2) this becomes

$$l(A_1\xi + A_2\eta + A_3\zeta) + m(B_1\xi + B_2\eta + B_3\zeta) + n(C_1\xi + C_2\eta + C_3\zeta) = 0. \quad (104.1)$$

This equation is linear in ξ, η and ζ. Conversely, a linear equation in ξ, η, ζ yields by (101.1) a linear equation in x, y and z. Hence a a linear equation in ξ, η, ζ always represents a stright line.

Two straight lines $\lambda_1\xi + \mu_1\eta + \nu_1\zeta = 0$ and $\lambda_2\xi + \mu_2\eta + \nu_2\zeta = 0$ clearly intersect in the point $(\mu_1\nu_2 - \mu_2\nu_1, \nu_1\lambda_2 - \nu_2\lambda_1, \lambda_1\mu_2 - \lambda_2\mu_1)$.

The equation of the straight line through P and Q is

$$\begin{vmatrix} \xi & \eta & \zeta \\ \xi_P & \eta_P & \zeta_P \\ \xi_Q & \eta_Q & \zeta_Q \end{vmatrix} = 0.$$

Further, the straight lines $\lambda_i\xi + \mu_i\eta + \nu_i\zeta = 0$ for $i = 1, 2, 3$ are concurrent if

$$\begin{vmatrix} \lambda_1 & \mu_1 & \nu_1 \\ \lambda_2 & \mu_2 & \nu_2 \\ \lambda_3 & \mu_3 & \nu_3 \end{vmatrix} = 0.$$

Example 3. Show that (104.1) cannot reduce to an identity. (Hint, $R \neq 0$.)

105. Pencil of lines

A pencil of straight lines is the dual of a range of points and so we may dualize immediately from §103 and obtain the results:

The pencil of lines determined by the lines $p \equiv (\lambda_p, \mu_p, \nu_p)$ and $q \equiv (\lambda_q, \mu_q, \nu_q)$ are given by

$$(\rho\lambda_p + \tau\lambda_q, \quad \rho\mu_p + \tau\mu_q, \quad \rho\nu_p + \tau\nu_q)$$

and the cross-ratio of the four lines corresponding to ρ_i, τ_i for $i = 1, 2, 3, 4$ is the cross-ratio of the four values of τ_i/ρ_i.

Further, the four lines $(\lambda_p, \mu_p, \nu_p)$, $(\lambda_q, \mu_q, \nu_q)$ and $(\rho\lambda_p \pm \tau\lambda_q, \rho\mu_p \pm \tau\mu_q, \rho\nu_p \pm \tau\nu_q)$ form a harmonic pencil.

Example 4. Prove that the four straight lines $y = 0$, $ax + hy = 0$ and $ax^2 + 2hxy + by^2 = 0$ form a harmonic pencil.

106. Quadrangle

The figure formed by four points P, Q, R and S, no three of which are collinear is called a **quadrangle** (Fig. 61). The six lines PR,

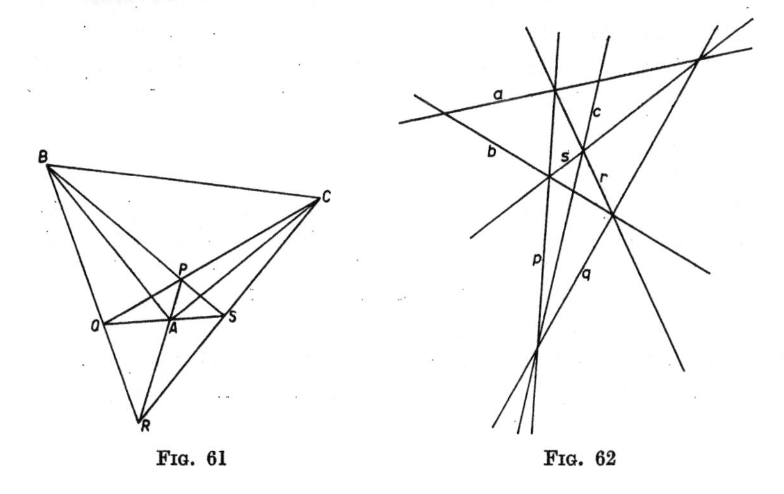

FIG. 61 FIG. 62

QS; PS, QR; PQ, SR are called pairs of **opposite sides,** and their points of intersection A, B, C are called the **diagonal points.** The triangle ABC is called the **diagonal point triangle.**

107. Quadrilateral

The **quadrilateral** is the figure dual to the quadrangle. It is formed by four lines p, q, r, s no three of which are concurrent (Fig. 62). The six points of intersection of p and r, q and s; p and s, q and r; p and q, s and r are called pairs of **opposite vertices,** and the lines a, b, c joining them are called the **diagonal lines.** The triangle abc is called the **diagonal line triangle.**

108. Arbitrary assignment of coordinates $(\pm 1, \pm 1, \pm 1)$

We wish to show that it is possible to assign the four sets of coordinates $(\pm 1, \pm 1, \pm 1)$ arbitrarily to any four points P, Q, R

and S no three of which are collinear. Select the diagonal point triangle (Fig. 61) ABC as the triangle of reference and P as the unit point $(1, 1, 1)$. (If PQ and SR are parallel, C is a point at infinity.) The equation of AP is $\eta - \zeta = 0$. Thus $R \equiv (\sigma, 1, 1)$ for some value of σ. Similarly $S \equiv (1, \tau, 1)$ and $Q \equiv (1, 1, \rho)$ for some values of τ and ρ. The points CSR are collinear and so

$$\begin{vmatrix} 0 & 0 & 1 \\ 1 & \tau & 1 \\ \sigma & 1 & 1 \end{vmatrix} = 0.$$

That is $\tau\sigma = 1$. Similarly, the collinearity of QAS and BQR yield $\rho\tau = 1$ and $\sigma\rho = 1$ respectively. Hence $\sigma^2\tau^2\rho^2 = 1$, from which $\sigma^2 = \tau^2 = \rho^2 = 1$. But Q, R, S do not coincide with P and so we cannot have $\sigma = 1, \tau = 1, \rho = 1$. Thus $\sigma = \tau = \rho = -1$. Accordingly $Q \equiv (1, 1, -1)$, $R \equiv (-1, 1, 1)$ and $S \equiv (1, -1, 1)$ as required.

Dually we can assign the four sets of coordinates $(\pm 1, \pm 1, \pm 1)$ arbitrarily to any four lines p, q, r and s, no three of which are collinear. In Fig. 62 we select the vertices of the diagonal line triangle abc as triangle of reference and the line p as the unit line $(1, 1, 1)$. (The reader is recommended to write out the proof in detail.)

Example 5. Show that a pair of opposite sides of a quadrangle are harmonically conjugate to the two sides of the diagonal point triangle which are concurrent with the given opposite sides.

State the dual theorem.

109. Conic

In virtue of (101.1) and (101.2) a quadratic equation in x, y, z transforms into a quadratic equation in ξ, η, ζ and conversely. Hence the quadratic equation

$$S(\xi, \eta, \zeta) \equiv S \equiv a\xi^2 + b\eta^2 + c\zeta^2 + 2f\eta\zeta + 2g\zeta\xi + 2h\xi\eta = 0$$

represents a conic.

The theory centred around Joachimsthal's quadratic equation follows as in §§68, 69 and 70. With the notation

$$S_P \equiv S(\xi_P, \eta_P, \zeta_P); \quad S_Q \equiv S(\xi_Q, \eta_Q, \zeta_Q);$$

$$\left.\begin{aligned} T_{PQ} = T_{QP} &\equiv (a\xi_P + h\eta_P + g\zeta_P)\xi_Q + (h\xi_P + b\eta_P + f\zeta_P)\eta_Q \\ &\quad + (g\xi_P + f\eta_P + c\zeta_P)\zeta_Q, \\ &\equiv (a\xi_Q + h\eta_Q + g\zeta_Q)\xi_P + (h\xi_Q + b\eta_Q + f\zeta_Q)\eta_P \\ &\quad + (g\xi_Q + f\eta_Q + c\zeta_Q)\zeta_P, \end{aligned}\right\} \tag{109.1}$$

$$\left.\begin{aligned} T_P \equiv (a\xi_P + h\eta_P + g\zeta_P)\xi + (h\xi_P + b\eta_P + f\zeta_P)\eta \\ + (g\xi_P + f\eta_P + c\zeta_P)\zeta, \\ \equiv (a\xi + h\eta + g\zeta)\xi_P + (h\xi + b\eta + f\zeta)\eta_P \\ + (g\xi + f\eta + c\zeta)\zeta_P, \end{aligned}\right\} \tag{109.2}$$

we list the following results:

(i) The points P and Q are conjugate if $T_{PQ} = 0$.

(ii) The polar of P is the straight line $T_P = 0$.

(iii) The tangent at P is $T_P = 0$.

(iv) The pair of tangents from P to the conic is $S_P S - T_P^2 = 0$.

(v) The line $\lambda\xi + \mu\eta + \nu\zeta = 0$ touches the conic if

$$\mathscr{A}\lambda^2 + \mathscr{B}\mu^2 + \mathscr{C}\nu^2 + 2\mathscr{F}\mu\nu + 2\mathscr{G}\nu\lambda + 2\mathscr{H}\lambda\mu = 0. \tag{109.3}$$

(vi) The pole of the line $\lambda\xi + \mu\eta + \nu\zeta = 0$ is at the point

$$(\mathscr{A}\lambda + \mathscr{H}\mu + \mathscr{G}\nu,\ \ \mathscr{H}\lambda + \mathscr{B}\mu + \mathscr{F}\nu,\ \ \mathscr{G}\lambda + \mathscr{F}\mu + \mathscr{C}\nu).$$

(vii) The conic degenerates to a line-pair if $\Delta = 0$.

The properties (e.g. centre, asymptotes, etc.) based on the behaviour of the conic with respect to the line at infinity cannot be discussed until the equation of the line at infinity is specified.

The line-equation of the conic is (109.3). It is left as an exercise for the reader to write down the dual results corresponding to (i)–(vii) of this section.

110. Special forms of equation of conic

In this section, we shall show that the equation of a conic can often be simplified by a judicious choice of the triangle of reference.

In all cases we start with the equation of the conic in its most general form

$$S \equiv a\xi^2 + b\eta^2 + c\zeta^2 + 2f\eta\zeta + 2g\zeta\xi + 2h\xi\eta = 0. \tag{110.1}$$

(1) Circumscribed conic

This refers to the case when the vertices of the triangle of reference lie on the conic (Fig. 63). Since $A \equiv (1, 0, 0)$ lies on (110.1) we have $a = 0$; similarly $b = c = 0$. Hence the circumscribed conic is represented by the equation

$$f\eta\zeta + g\zeta\xi + h\xi\eta = 0. \tag{110.2}$$

We could choose the unit point so that $f = g = h$ and the conic simplifies further to

$$\eta\zeta + \zeta\xi + \xi\eta = 0.$$

In general, it is advisable to retain (110.2) in tackling problems as this allows any other relevant point occurring in the problem to be made the unit point. Further all equations remain homogeneous in f, g and h; if an equation is obtained which is not homogeneous then an error has been made.

Example 6. The triangle ABC is inscribed in a conic. The tangents at A, B and C intersect the opposite sides BC, CA and AB in D, E and F respectively. Show that the points D, E and F are collinear.

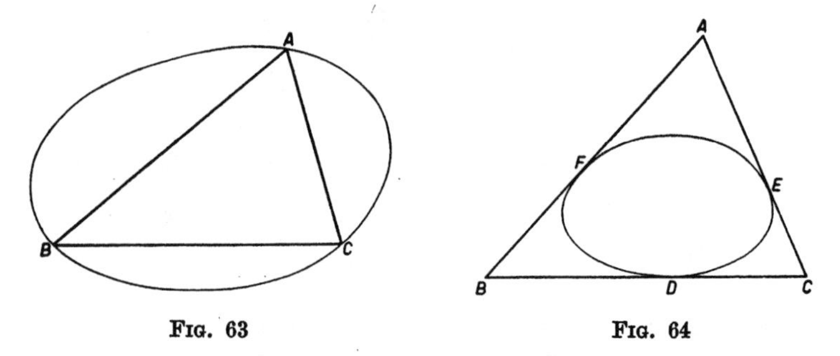

FIG. 63 FIG. 64

(2) Inscribed conic

This refers to the case when the sides of the triangle of reference touch the conic (Fig. 64). Since $\xi = 0$ is a tangent to the conic (110.1) the equation $b\eta^2 + c\zeta^2 + 2f\eta\zeta = 0$ has coincident roots. Thus $bc = f^2$. Similarly $ca = g^2$ and $ab = h^2$. Hence a, b, c are either all positive or all negative and there is no loss in generality if we select them to be positive. Let us write $a = \alpha^2$, $b = \beta^2$, $c = \gamma^2$; then the inscribed conic has the equation

$$\alpha^2\xi^2 + \beta^2\eta^2 + \gamma^2\zeta^2 \pm 2\beta\gamma\eta\zeta \pm 2\gamma\alpha\zeta\xi \pm 2\alpha\beta\xi\eta = 0. \quad (110.3)$$

We cannot allow the choices of sign which make (110.3) a perfect square since the conic is non-degenerate. Accordingly we have four possibilities, namely $-, -, -$ or $+, +, -$ or $+, -, +$ or $-, +, +$. These correspond to the fact that the conic may be inscribed or escribed to the triangle of reference.

The points of contact of the conic with the sides of the triangle of reference are easily calculated to be $D \equiv (0, \gamma, \pm\beta)$, $E \equiv (\gamma, 0, \pm\alpha)$ and $F \equiv (\beta, \pm\alpha, 0)$.

If we select the unit point inside the triangle of reference, the points of contact must be $(0, \gamma, \beta)$, etc., and so the inscribed conic has the equation

$$\alpha^2\xi^2 + \beta^2\eta^2 + \gamma^2\zeta^2 - 2\beta\gamma\eta\zeta - 2\gamma\alpha\zeta\xi - 2\alpha\beta\xi\eta = 0. \quad (110.4)$$

Example 7. A triangle is circumscribed to a conic. Show that the three lines joining the vertices to the points of contact of the opposite sides are concurrent.

(3) Conic with self-conjugate triangle

Choose the self-conjugate triangle as the triangle of reference (Fig. 65). The points $A \equiv (1, 0, 0)$ and $B \equiv (0, 1, 0)$ are conjugate

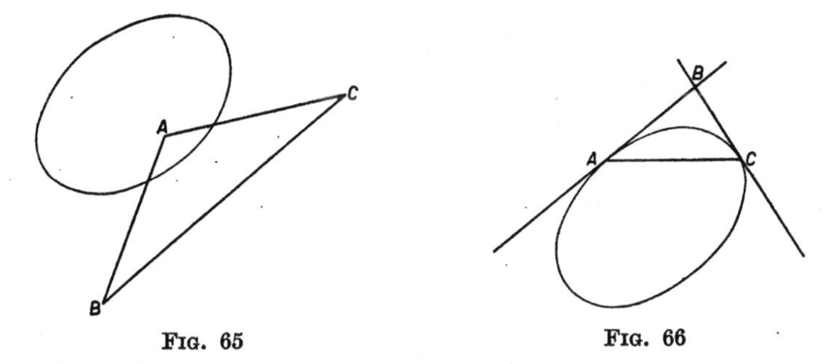

FIG. 65 FIG. 66

with respect to (110.1). Hence by §109 (i) we have $h = 0$; similarly $f = g = 0$. Consequently the conic is represented by

$$a\xi^2 + b\eta^2 + c\zeta^2 = 0. \quad (110.5)$$

Example 8. If two triangles have a common self-conjugate triangle, prove that their six vertices lie on another conic.

(4) Conic touching two lines

Select the two lines and the lines joining the points of contact as the triangle of reference (Fig. 66). Since the points A and C lie on the conic (110.1), we have $a = c = 0$. From the fact that $\xi = 0$ and $\zeta = 0$ are tangents it follows that $bc - f^2 = 0$ and $ab - h^2 = 0$. Thus $f = h = 0$. Accordingly the conic simplifies to $b\eta^2 + 2g\zeta\xi = 0$. If we take the unit point on the conic we may simplify further to

$$\eta^2 - \zeta\xi = 0.$$

This form of the equation lends itself to the parametric representation of the conic as the locus of the points $(t^2, t, 1)$ as t varies.

Example 9. Show that the equation of the chord joining the points $(t_1^2, t_1, 1)$ and $(t_2^2, t_2, 1)$ on the conic $\eta^2 - \zeta\xi = 0$ is $x - (t_1 + t_2)y + t_1t_2z = 0$. Hence, the equation of the tangent at the point t is $x - 2ty + t^2z = 0$.

Example 10. The points $P_1 \equiv (t_1^2, t_1, 1)$ and $P_2 \equiv (t_2^2, t_2, 1)$ on the conic $\eta^2 - \zeta\xi = 0$ are said to be in homographic correspondence if t_1, t_2 are connected by a homographic correspondence.

Show that the lines joining pairs of points in involution on a conic are concurrent.

The **dual results** in line-coordinates are as follows:

(1) The inscribed conic has the equation

$$f\mu\nu + g\nu\lambda + h\lambda\mu = 0.$$

(2) The circumscribed conic is

$$\alpha^2\lambda^2 + \beta^2\mu^2 + \gamma^2\nu^2 \pm 2\beta\gamma\mu\nu \pm 2\gamma\alpha\nu\lambda \pm 2\alpha\beta\lambda\mu = 0.$$

where the four combinations of signs listed above are possible.

(3) The conic with self-conjugate triangle of reference is

$$a\lambda^2 + b\mu^2 + c\nu^2 = 0.$$

(4) The conic touching two sides of a triangle of reference at two of its vertices has the equation

$$b\mu^2 + 2g\nu\lambda = 0,$$

or

$$\mu^2 - \nu\lambda = 0.$$

Example 11. Prove the results of Examples 6 and 7, using line coordinates

Example 12. Dualize the result of Example 8 and give the proof by line-coordinates.

Example 13. Dualize the definition and theorem of Example 10 and prove by the method of line-coordinates.

111. Pencils of conics

In Chapter 11 we already introduced the various types of pencils of conics. Now we find the corresponding equations using generalized homogeneous coordinates.

(1) Pencil with four distinct base points

In §108 we saw that we can assign the coordinates $(\pm 1, \pm 1, \pm 1)$ to the four points A, B, C, D (Fig. 41), by choosing the diagonal point triangle as the triangle of reference. Since these points lie on the conic (110.1) we have

$$\begin{aligned} a + b + c + 2f + 2g + 2h &= 0, \\ a + b + c + 2f - 2g - 2h &= 0, \\ a + b + c - 2f + 2g - 2h &= 0, \\ a + b + c - 2f - 2g + 2h &= 0, \end{aligned}$$

from which

$$a + b + c = f = g = h = 0.$$

Hence the pencil of conics is represented by

$$a\xi^2 + b\eta^2 + c\zeta^2 = 0, \tag{111.1}$$

subject to

$$a + b + c = 0. \tag{111.2}$$

In problems, it is often convenient to choose the pencil as

$$\xi^2 + \rho\eta^2 - (1 + \rho)\zeta^2 = 0, \tag{111.3}$$

the condition (111.2) being satisfied automatically.

Example 14. Prove that the locus of the poles of a fixed straight line with respect to a pencil of conics through four fixed points is a conic which passes through the diagonal point triangle of the quadrangle formed by the four base points.

(2) Single contact

Choose ABC, Fig. 42, as the triangle of reference. All the conics of the pencil circumscribe the triangle of reference and so the pencil is included in

$$f\eta\zeta + g\zeta\xi + h\xi\eta = 0.$$

But all conics of the pencil have a common tangent $g\zeta + h\eta = 0$ at A. Thus the ratio $g:h$ remains constant for all conics of the pencil. Hence the pencil of conics with single contact at A is represented by

$$\rho\eta\zeta + g\zeta\xi + h\xi\eta = 0,$$

where ρ varies from conic to conic of the pencil.

In problems, it is often convenient to choose the unit point so that $g = h$. Then the pencil can be represented by

$$\rho\eta\zeta + \zeta\xi + \xi\eta = 0.$$

Example 15. Prove that the locus of the poles of a fixed straight line with respect to a pencil of conics which have single contact at A and cut in the points B and C is a conic which passes through the following points: (a) A, (b) the point of intersection of BC with the common tangent at A, (c) the harmonic conjugate with respect to B and C of the point of intersection of the fixed line with BC.

(3) Double contact

From §110 (4) we see that the pencil of conics having double contact at A and C (Fig. 65) is

$$b\eta^2 + 2g\zeta\xi = 0. \tag{111.6}$$

It is sometimes convenient to write this equation

$$\rho\eta^2 + 2\zeta\xi = 0 \tag{111.7}$$

Example 16. Prove that the locus of the poles of a fixed straight line with respect to a pencil of conics which have double contact at A and C is the straight line joining the point of intersection of the common tangents at A and C to the harmonic conjugate with respect to A and C of the point of intersection of the fixed line and AC.

(4) Three-point contact

Consider the pencil of conics having three-point contact at A and (Fig. 67) passing through B. Take the triangle of reference to be ABC, where C is any point on the tangent at A. Since A and B

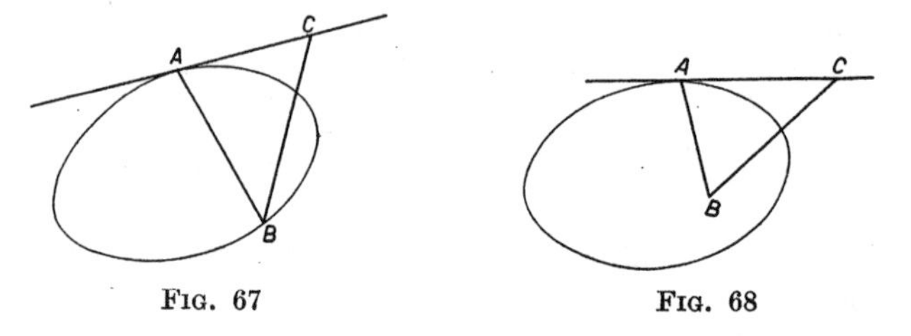

FIG. 67   FIG. 68

lie on the conic (110.1) we have $a = b = 0$. The tangent at A is $\eta = 0$ and so $g = 0$. Thus the pencil of conics is included in

$$c\zeta^2 + 2f\eta\zeta + 2h\xi\eta = 0. \tag{111.8}$$

The degenerate conics of this pencil are AC and AB, taken three times; that is, $\eta\zeta = 0$. This can only be obtained from (111.6) if we allow f to vary and then the degeneracy corresponds to infinite f. Accordingly the pencil is given by

$$c\zeta^2 + 2\rho\eta\zeta + 2h\xi\eta = 0, \tag{111.9}$$

where ρ varies from conic to conic of the pencil.

Example 17. Prove that the locus of the poles of a fixed line with respect to a pencil of conics having three-point contact at A and passing through a common point B is a conic which passes through A and the harmonic conjugate with respect to A and B of the point of intersection of the fixed straight line and AB.

(5) Four-point contact

In Fig. 68, the pencil of conics are to have four-point contact at A. Select the triangle ABC of reference, where C is any point on the common tangent at A and B is any point of the plane. Since A lies on the conic we have $a = 0$. The common tangent at A is $\eta = 0$ and so $g = 0$. Thus the pencil of conics is included in

$$b\eta^2 + c\zeta^2 + 2f\eta\zeta + 2h\zeta\eta = 0.$$

The degeneracies of this pencil are the coincident tangents at A, taken three times; that is $\eta^2 = 0$. Hence the pencil of conics is represented by

$$\rho\eta^2 + c\zeta^2 + 2f\eta\zeta + 2h\xi\eta = 0, \tag{111.10}$$

where ρ varies from conic to conic of the pencil.

Example 18. Prove that the locus of the poles of a fixed straight line with respect to a pencil of conics having four-point contact at A is a straight line through A.

Miscellaneous examples

1. Two pairs of opposite vertices of a quadrilateral are each conjugate with respect to a conic. Prove that the third pair is also conjugate with respect to the conic.

2. A pencil of conics has single contact at A and passes through the points B and C. A fixed line through A cuts a conic of the pencil at P. The tangent to this conic at P intersects AB and CA in L and M respectively. The straight lines BM and CL intersect in Q. Show that PQ passes through a fixed point.

3. A conic S touches the sides of a triangle ABC at D, E and F respectively. D' is the harmonic conjugate to D with respect to B and C. The tangent from D' to the conic S, other than the straight line BC, cuts AC and AB in L and M respectively and touches the conic at X. The straight lines BL and CM intersect in P. Show that A, D, P and X are collinear.

Further the points Q and R are defined in a similar manner to P. Prove that if the locus of R is a straight line, then PQ envelopes a conic, which is inscribed in the triangle ABC.

4. Show that the envelope of the straight line, joining points in homographic correspondence on a conic, is a conic having double contact with the original conic.

5. C is any point on the common tangent at A to all conics of a pencil having three-point contact at A. The other tangent through C to a conic of the pencil has contact at D. Show that the locus of the points D is a conic touching CA and CB.

Further the straight line CB intersects the conic again at E. Show that the envelope of the straight line DE is another conic touching CA and CB.

6. A pencil of conics has four-point contact at A, and C is a point on the common tangent at A. B is any point not on AC and the conic S_1 of the pencil touches BC at L whilst the conic S_2 of the pencil which passes through the point B intersects BC at M. Show that B, M divide L, C harmonically.

7. A conic Σ circumscribes the quadrangle $PQRS$. A conic Σ_1 is drawn to have double contact with Σ at P and Q and to pass through the point A of intersection of PR and QS. Another conic Σ_2 is also drawn through A, having double contact with Σ at R and S. Show that the conics Σ_1 and Σ_2 have single contact at A.

B and C are the points of intersection of QR, PS and PQ, RS respectively. Show that the two common tangents to Σ_1 and Σ_2, which do not pass through A, separate BA and BC harmonically.

8. If a hexagon is inscribed in a conic, the points of intersection of pairs of opposite sides are collinear. This theorem is known as *Pascal's Theorem.* State the dual theorem, known as *Brianchon's Theorem.*

9. If there is one triangle inscribed in a conic S and circumscribed about a conic Σ, then an infinite number of triangles can be inscribed in S and circumscribed to Σ.

APPENDIX

SOLUTIONS

Chapter II. Miscellaneous examples

1. Let the line-pair $ax^2 + 2hxy + by^2 = 0$ be $l_1x + m_1y = 0$ and $l_2x + m_2y = 0$. Then $\dfrac{l_1l_2}{a} = \dfrac{l_1m_2 + l_2m_1}{2h} = \dfrac{m_1m_2}{b}$. The perpendicular line-pair is $(m_1x - l_1y)(m_2x - l_2y) = 0$. Hence, etc.

2. Let the mirror image be at (γ, δ). Then two linear equations in γ and δ follow from the facts that the mid-point $(\frac{1}{2}(\gamma + \alpha), \frac{1}{2}(\delta + \beta)$ lies on the given straight line and the gradient $(\delta - \beta)/(\gamma - \alpha)$ is perpendicular to $- l/m$.

3. Let (x_1, y_1) and (x_2, y_2) be the points of intersection of $lx + my + n = 0$ and $ax^2 + 2hxy + by^2 = 0$. The required area is $\frac{1}{2}(x_1y_2 - x_2y_1)$. From $lx_1 + my_1 + n = 0$ and $lx_2 + my_2 + n = 0$, we have $m(x_1y_2 - x_2y_1) = n(x_2 - x_1)$. Further the elimination of y yields the quadratic $(am^2 - 2hlm + bl^2)x^2 + 2n(bl - hm)x + bn^2 = 0$. Hence $x_2 - x_1$ can be calculated, etc.

4. Let the line-pair be $(l_1x + m_1y)(l_2x + m_2y) = 0$. The pair of bisectors is given by $(l_1x + m_1y)/\sqrt{l_1^2 + m_1^2} = \pm\, (l_2x + m_2y)/\sqrt{l_2^2 + m_2^2}$. That is, $(l_2^2 + m_2^2)(l_1x + m_1y)^2 - (l_1^2 + m_1^2)(l_2x + m_2y)^2 = 0$, which reduces on division by $m_2l_1 - m_1l_2$ $(\neq 0)$ to $(m_2l_1 + m_1l_2)(x^2 - y^2) - 2(l_1l_2 - m_1m_2)xy = 0$. Hence, etc.

5. The results follows immediately from (17.1).

7. The equation of the line-pair may be written

$$ax + hy + g = \pm \sqrt{(h^2 - ab)y^2 + 2(hg - af)y + g^2 - ac}.$$

The condition $\Delta = 0$ ensures that the quadratic in y under the radical sign is a perfect square. Thus by addition, the point of intersection of the line-pair lies on $ax + hy + g = 0$. Similarly it lies in $hx + by + f = 0$. Hence, etc.

8. By Exercise 4, the equation of the bisectors of the angles is $h(x^2 - y^2) - (a - b)xy = 0$ and its gradients satisfy the quadratic equation $h(1 - g^2) - (a - b)g = 0$. One of these bisectors is perpendicular to $lx + my + n = 0$. Thus m/l satisfies the quadratic in g. Hence, etc.

Chapter III. Miscellaneous examples

1. Select the line joining the centres and its perpendicular bisector as x-axis and y-axis respectively. Let $2a$ be distance between centres and let the equations of the circles be $(x - a)^2 + y^2 = r_1^2$ and $(x + a)^2 + y^2 = r_2^2$. The polars of (α, β) are $\alpha x + \beta y - a(x + \alpha) + a^2 - r_1^2 = 0$ and $\alpha x + \beta y + a(x + \alpha) + a^2 - r_2^2 = 0$. The orthogonality condition is $(\alpha - a)(\alpha + a) + \beta^2 = 0$. Hence the locus of (α, β) is the circle $x^2 + y^2 = a^2$ as required.

2. The equation of any circle through the points of intersection of $x^2 + y^2 + 2g_1x + 2f_1y + c_1 = 0$ and $l_1x + m_1y + n_1 = 0$ is $x^2 + y^2 + 2g_1x + 2f_1y + c_1 + \lambda_1(l_1x + m_1y + n_1) = 0$. This must be identical with $x^2 + y^2 + 2g_2x + 2f_2y + c_2 + \lambda_2(l_2x + m_2y + n_2) = 0$ for some values of λ_1 and λ_2. Comparing coefficients, we obtain three equations. The elimination of λ_1 and λ_2 yields the required result.

3. From a diagram it is clear that the line joining centres and the two radii form a right-angled triangle whose hypotenuse is the radius of the first circle. Hence $g_1^2 + f_1^2 - c_1 = (g_1 - g_2)^2 + (f_1 - f_2)^2 + g_2^2 + f_2^2 - c_2$, etc.

4. The coaxal system formed by the two circles is $x^2 + y^2 + 18x - 2y - 15 + \lambda(x^2 + y^2 - 6x - 4y - 5) = 0$. The centre of the circle is at $((3\lambda - 9)/(\lambda + 1), (2\lambda + 1)/(\lambda + 1))$. For circle required, this centre lies on the radical axis $12x + y - 5 = 0$. Hence $\lambda = 112/33$, etc. (Alternatively write down an expression for the square of the radius of circle and use the fact that this quantity is a minimum for the required circle.)

5. The sum of the squares of the tangents from P and Q to the circle $S = 0$ less $(PQ)^2$ is $x_P^2 + y_P^2 + 2gx_P + 2fy_P + c + x_Q^2 + y_Q^2 + 2gx_Q + 2fy_Q + c - (x_P - x_Q)^2 - (y_P - y_Q)^2 = 2[x_Px_Q + y_Py_Q + g(x_P + x_Q) + f(y_P + y_Q) + c]$. Hence, etc.

6. The straight line $y = mx$ intersects the circle $x^2 + y^2 + 2gx + c = 0$, where $(1 + m^2)x^2 + 2gx + c = 0$. If (α, β) is the mid-point of the chord, then $\alpha = -g/(1 + m^2)$, $\beta = m\alpha$. Eliminate m and using running coordinates, the required result follows.

8. The circumcircle has its centre at $(0, \mu)$ and radius $\sqrt{\lambda^2 + \mu^2}$. Hence its equation is $x^2 + (y - \mu)^2 = \lambda^2 + \mu^2$. The point (α, β) lies on this circle, hence μ is obtained, etc.

The nine-points circle passes through $(0, 0)$ and $(\alpha, 0)$. Hence its centre is at $\left(\frac{\alpha}{2}, \mu\right)$ and so its equation is $\left(x - \frac{\alpha}{2}\right)^2 + (y - \mu)^2 = \frac{\alpha^2}{4} + \mu^2$. The value of μ is obtained from the fact that $(\frac{1}{2}(\alpha + \lambda), \frac{1}{2}\beta)$ lies on the circle. Hence, etc.

9. The equation of all circles in first quadrant which touch both coordinate axes is $(x - \mu)^2 + (y - \mu)^2 = \mu^2$ $(\mu > 0)$. Hence, etc.

10. The square of the radius of a circle of coaxal system is $\{g^2 + \lambda^2 f^2 - (1 + \lambda)(c + \lambda c')\}/(1 + \lambda)^2 = \{\lambda^2(f^2 - c') - \lambda(c + c') + g^2 - c\}/(1 + \lambda)^2$. The limiting points are real if the quadratic $\lambda^2(f^2 - c') - \lambda(c + c') + g^2 - c = 0$ has real roots. Hence, etc.

Chapter IV

7. Compare $lx + my + n = 0$ and the normal $\frac{x}{b}\sin\theta - \frac{y}{a}\cos\theta = \frac{a^2 - b^2}{ab}$ $\sin\theta\cos\theta$. From the resulting two equations eliminate θ. Hence, etc.

8. The required line has the equation $x - x_A + \lambda(y - y_A) = 0$. We obtain λ from the fact that the line $x + \lambda y = 0$ is conjugate to the line $y_Ax - x_Ay = 0$. Hence, etc.

13. The required condition is $\begin{vmatrix} a\cos\theta & b\sin\theta & 1 \\ a\cos\phi & b\sin\varphi & 1 \\ ae & 0 & 1 \end{vmatrix} = 0$, which reduces to

$\sin(\theta - \phi) = e(\sin\theta - \sin\phi)$. That is $\cos\frac{1}{2}(\theta - \phi) = e\cos\frac{1}{2}(\theta + \phi)$. The negative sign turns up if we use the other focus at $(-ae, 0)$.

14. Let Q and R have eccentric angles θ and φ respectively. Use the previous result, together with Example 5, which yields $a\cos\frac{1}{2}(\theta + \phi)/\cos\frac{1}{2}(\theta - \phi) = a/e$.

15. Let P, A and B have eccentric angles θ, α and β respectively.

Then from Example 13, $\cos\ \frac{1}{2}(\theta - \alpha) = e \cos \frac{1}{2}(\theta + \alpha)$ and $\cos \frac{1}{2}(\theta - \beta) = -e \cos \frac{1}{2}(\theta + \beta)$. Verify that point of intersection of tangents at A and B (see Example 5) lies on the normal at P.

Miscellaneous examples

2. Let the pole of the normal $\frac{x}{b} \sin \theta - \frac{y}{a} \cos \theta = \frac{a^2 - b^2}{ab} \sin \theta \cos \theta$ be at (α, β). Then this normal is also $\frac{\alpha x}{a^2} + \frac{\beta y}{b^2} = 1$. Compare coefficients and from resulting two equations eliminate θ to obtain the required result.

3. Let the ends of the chord have eccentric angles θ and φ. Then the straight lines $y/x = b \sin \theta/a \cos \theta$ and $y/x = b \sin \phi/a \cos \phi$ are orthogonal. That is, $a^2 \cos \theta \cos \phi + b^2 \sin \theta \sin \phi = 0$. Eliminate θ and ϕ between this equation and (see Example 5, page 38) $x = a \cos \frac{1}{2}(\theta + \phi)/\cos \frac{1}{2}(\theta - \phi)$, $y = b \sin \frac{1}{2}(\theta + \phi)/\cos \frac{1}{2}(\theta - \phi)$.

4. Let the gradients of the conjugate diameters be m and m', where $mm' = -b^2/a^2$. The straight lines are $y = m(x - ae)$ and $y = m'(x + ae)$. Hence the required locus is $y^2 = -b^2(x^2 - a^2e^2)/a^2$. That is $\frac{x^2}{a^2e^2} + \frac{y^2}{b^2e^2} = 1$. Hence, etc.

5. Let A, B and P have eccentric angles α, β and θ respectively. Then $\{-b \cos \frac{1}{2}(\alpha + \theta)/a \sin \frac{1}{2}(\alpha + \theta)\}\{-b \cos \frac{1}{2}(\beta + \theta)/a \sin \frac{1}{2}(\beta + \theta)\} = -(b^2 + c^2)/(a^2 + c^2)$. That is, $\cos \frac{1}{2}(\alpha + \theta) \cos \frac{1}{2}(\beta + \theta)/\sin \frac{1}{2}(\alpha + \theta) \sin \frac{1}{2}(\beta + \theta) = -a^2(b^2 + c^2)/b^2(a^2 + c^2)$, from which $\cos \frac{1}{2}(\alpha + \beta + 2\theta)/\cos \frac{1}{2}(\alpha - \beta) = \{2a^2b^2 + b^2c^2 + c^2a^2\}/c^2(a^2 - b^2)$. Hence $\cos \frac{1}{2}(\alpha + \beta) \cos \theta - \sin \frac{1}{2}(\alpha + \beta) \sin \theta = (2a^2b^2 + b^2c^2 + c^2a^2) \cos \frac{1}{2}(\alpha - \beta)/c^2(a^2 - b^2)$. Hence the chord $\frac{x}{a} \cos \frac{1}{2}(\alpha + \beta) + \frac{y}{b} \sin \frac{1}{2}(\alpha + \beta) = \cos \frac{1}{2}(\alpha - \beta)$ passes through the fixed point $(ac^2(a^2 - b^2) \cos \theta/(2a^2b^2 + b^2c^2 + c^2a^2),\ -bc^2(a^2 - b^2) \sin \theta/(2a^2b^2 + b^2c^2 + c^2a^2))$.

7. The equation of the circle on PF as diameter is $(x - ae)(x - a \cos \theta) + y(y - b \sin \theta) = 0$. Verify that the given point is on both circles and that the tangents at this point are identical.

8. The result follows immediately from Example 6, page 38.

Chapter V

7. Let P and Q be the points α and β respectively. Then the equation of the chord PQ is given by (39.2). Hence the abscissae of R and S are $a \cos \frac{1}{2}(\alpha + \beta)/\{\cos \frac{1}{2}(\alpha - \beta) \pm \sin \frac{1}{2}(\alpha + \beta)\}$. Thus the abscissa of the mid-point of RS is $a(\cos \alpha + \cos \beta)/2 \cos \alpha \cos \beta$. This is also the abscissa of the mid-point of PQ. Similarly for the ordinates. Hence the mid-points of PQ and RS are identical. Hence, etc.

11. Let P, Q, R and S be the points t_1, t_2, t_3 and t_4 on the rectangular hyperbola. The gradients of PQ and RS are $-1/t_1t_2$ and $-1/t_3t_4$. Hence $t_1t_2t_3t_4 = -1$. The gradient of OP is $1/t_1^2$. Hence, etc.

Miscellaneous examples

1. Method similar to that in §35.

2. Method similar to that in §36.

3. The point $(ct, c/t)$ is on the circle $x^2 + y^2 + 2gx + 2fy + k = 0$ if $c^2t^4 + 2gct^3 + kt^2 + 2fct + c^2 = 0$. Let t_1, t_2, t_3 and t_4 be the parameters of the four points, then $\Sigma t_1 = -2g/c$, $\Sigma t_1t_2t_3 = -2f/c$ and $t_1t_2t_3t_4 = 1$. The centre of the circle is at $(-g, -f)$, whilst the centroid of the four points is at $\left(\frac{c}{4}\Sigma t_1, \frac{c}{4}\Sigma 1/t_1\right)$. But $\Sigma 1/t_1 = \Sigma t_1t_2t_3$ since $t_1t_2t_3t_4 = 1$. Hence, etc.

The sum of the squares of the distances from the centre is $c^2\Sigma t_1^2 + c^2\Sigma(1/t_1^2)$. But $\Sigma t_1^2 = (\Sigma t_1)^2 - 2\Sigma t_1t_2$ and $\Sigma(1/t_1)^2 = \Sigma t_1^2t_2^2t_3^2 = (\Sigma t_1t_2t_3)^2 - 2\Sigma t_1t_2$. Hence, etc.

4. The equation $\frac{x^2}{a^2} - \frac{y^2}{b^2} - 1 + (lx + my + n)(\lambda x + \mu y + \rho) = 0$ represents a circle through L and M if $1/a^2 + l\lambda = -1/b^2 + m\mu$ and $l\mu + m\lambda = 0$. It is the circle on LM as diameter if its centre lies on $lx + my + n = 0$. Hence we have three equations for λ, μ and ρ, etc.

5. Let the vertices of the triangle be at t_1, t_2 and t_3. The orthocentre is the point of intersection of the straight lines $t_1t_2x - y = ct_1t_2t_3 - c/t_3$ and $t_2t_3x - y = ct_1t_2t_3 - c/t_1$. Hence the orthocentre is at $(-c/t_1t_2t_3, -ct_1t_2t_3)$ and so lies on hyperbola as required.

6. Let A and A' be the points θ and φ. The tangent at A is $\frac{x}{a}\sec\theta - \frac{y}{b}\tan\theta = 1$. Hence $B \equiv (a\cos\theta/(1 - \sin\theta), b\cos\theta/(1 - \sin\theta))$. Similarly for B'. Hence the mid-point of BB' has the abscissa $a\{\cos\theta + \cos\varphi - \sin(\theta + \varphi)\}/2(1 - \sin\theta)(1 - \sin\varphi) = \cos\tfrac{1}{2}(\theta + \varphi)/\{\cos\tfrac{1}{2}(\theta - \varphi) - \sin\tfrac{1}{2}(\theta + \varphi)\}$. Hence, etc.

7. The normal at the point t is $t^3x - ty + c(1 - t^4) = 0$. If this passes through the point u on the hyperbola, $t^3u - t/u + 1 - t^4 = 0$. Removing the factor $t - u$, the result is $t^3u + 1 = 0$. Let t_1, t_2, t_3 be the parameters of A, B and C respectively. Then $\Sigma t_1 = 0$, $\Sigma t_1t_2 = 0$. Hence, etc. (Note that two of the points A, B and C are complex.)

8. The normal at P, the point θ, is $\frac{x}{b}\sin\theta + \frac{y}{a} = \frac{a^2 + b^2}{ab}\tan\theta$. Thus $L \equiv ((a^2 + b^2)\sec\theta/a, 0)$ and $M \equiv (0, (a^2 + b^2)\tan\theta/b)$, and so the mid-point of LM is $((a^2 + b^2)\sec\theta/2a, (a^2 + b^2)\tan\theta/2b)$ and its locus accordingly is $a^2x^2 - b^2y^2 = (a^2 + b^2)/4$. Hence, etc.

9. The normals at the points α, β intersect in the point $\left(\frac{a^2 + b^2}{a}\cos\tfrac{1}{2}(\alpha - \beta)\sec\tfrac{1}{2}(\alpha + \beta)\sec\alpha\sec\beta, -\frac{a^2 + b^2}{b}\tan\alpha\tan\beta\tan\tfrac{1}{2}(\alpha + \beta)\right)$. Let $\alpha \to \beta$, etc.

10. Obtain the coordinates of the points of intersection of the tangents at θ_1, θ_2 and θ_3, θ_4 respectively. (Example 3, page 48.) Then determine the coordinates of the mid-point of one of the diagonals of the circumscribed quadrilateral. Further verify that the gradient of the line joining the origin to this point is $2b\sin s/a\Sigma\sin(s - \theta_1)$, where $2s = \theta_1 + \theta_2 + \theta_3 + \theta_4$. From the symmetry of this gradient, the result follows.

11. If $P \equiv (a\sec\theta, b\tan\theta)$, the diameter conjugate to OP is $y/x = b\sec\theta/a\tan\theta$, and so $Q \equiv (a\tan\theta, b\sec\theta)$. The normals at P and Q are $\frac{x}{b}\sec\theta + \frac{y}{a}\tan\theta = \frac{a^2 + b^2}{ab}\sec\theta\tan\theta$ and $\frac{x}{b}\tan\theta + \frac{y}{a}\sec\theta = \frac{a^2 + b^2}{ab}\sec\theta\tan\theta$. Hence $ax - by = 0$. The other line $ax + by = 0$ is obtained if Q is on the second branch of the conjugate hyperbola.

Chapter VI. Miscellaneous examples

1. The point $(at^2, 2at)$ lies on the circle $x^2 + y^2 + 2gx + 2fy + c = 0$ if $a^2t^4 + (4a^2 + 2ga)t^2 + 4aft + c = 0$. Hence $\Sigma t_1 = 0$, etc.

2. The pair of tangents from (α, β) has the equation $(\beta^2 - 4a\alpha)(y^2 - 4ax) - (2ax - \beta y + 2a\alpha)^2 = 0$. Hence $\tan^2 \theta = (\beta^2 - 4a\alpha)/(a + \alpha)^2$, etc.

3. Let the ends of a chord have parameters t_1, t_2. The orthogonality condition is $t_1t_2 = -4$. The mid-point is $(\frac{1}{2}a(t_1^2 + t_2^2), a(t_1 + t_2))$. Hence, etc.

4. Let the ends of a focal chord have parameters t_1, t_2. Then $t_1t_2 = -1$. Thus the equation of the circle on the focal chord as diameter is $(x - at_1^2)(x - a/t_1^2) + (y - 2at_1)(y + 2a/t_1) = 0$, which simplifies to $x^2 + y^2 - a(1/t_1^2 + t_1^2)x + 2a(1/t_1 - t_1)y - 3a^2 = 0$. Hence, etc.

5. See Example 6, page 56.

6. The orthocentre of the triangle whose vertices are at t_1, t_2 and t_3 is the point of intersection of the straight lines $t_2x + y = at_1t_2t_3 + a(t_1 + t_3)$ and $t_3x + y = at_1t_2t_3 + a(t_1 + t_2)$. Hence, etc.

7. The normal at the point t passes through the fixed point (α, β) if $at^3 + (2a - \alpha)t - \beta = 0$. Hence $t_1 + t_2 + t_3 = 0$ if t_1, t_2, t_3 are the parameters of A, B and C. Hence, etc.

9. Let the vertices of the triangle ABC inscribed in $y^2 = 4ax$ have the parameters t_1, t_2, t_3. The conditions that the three sides touch $x^2 = 4by$ are $t_1t_2(t_1 + t_2) = t_2t_3(t_2 + t_3) = t_3t_1(t_3 + t_1) = -2b/a$. These are equivalent to the *two* equations $t_1 + t_2 + t_3 = 0$ and $t_1t_2t_3 = 2b/a$. Hence an infinite number of solutions exist.

Let t_4 be the parameter of the fourth point of intersection of the circumcircle of ABC and the parabola $y^2 = 4ax$. From Example 1 we have $t_1 + t_2 + t_3 + t_4 = 0$. Hence $t_4 = 0$, etc.

10. The normal at $(at^2, 2at)$ passes through $P \equiv (au^2, 2au)$ if $tu^2 + 2u - t^3 - 2t = 0$. Remove the factor $t - u$ and we find that the parameters t_1, t_2 of Q and R are the roots of $t^2 + ut + 2 = 0$. Hence $t_1t_2 = 2$, etc.

Chapter VII. Miscellaneous examples

2. Translate to parallel axes through (1, 2) and the equations of the chord and hyperbola become $2x + 3y = -7$ and $x^2 - y^2 + 2x - 4y - 4 = 0$. The lines joining the new origin to the points of intersection of the chord and the hyperbola (see §17) are given by $49(x^2 - y^2) - 7(2x + 3y)(2x-4y) - 4(2x + 3y)^2 = 0$. That is, $5x^2 - 34xy - y^2 = 0$. Hence, etc.

3. The relative invariants of the two curves are $I = 1$; $J = \cos^2 \theta \sin^2\theta (1 - \lambda^2)$; $\Delta = -J$ and $I = 1$; $J = -\sec^2 \theta \tan^2 \theta$; $\Delta = -J$. Equating the absolute invariant J/I^2 we have $\cos^2 \theta \sin^2 \theta(1 - \lambda^2) = -\sec^2 \theta \tan^2 \theta$, etc. (Note that here the same result is obtained by equating a second absolute invariant. If this were not so, the two curves could not be identical.)

4. The relative invariants of the two curves are $I = -1$; $J = -6$; $\Delta = 42$ and $I = 1 + \lambda$; $J = \lambda - 25$; $\Delta = 3\lambda\mu - 49\lambda - 75\mu + 649$. From I^2/J we have $6\lambda^2 + 13\lambda - 19 = 0$. Hence $\lambda = 1$ or $-19/6$. The corresponding values of μ are obtained by considering a second absolute invariant, say Δ/I^3.

Chapter VIII. Miscellaneous examples

2. The intersection of the conic $ax^2 + 2hxy + by^2 + 2gx + 2fy + c = 0$ and the y-axis yields the equation $ax^2 + 2gx + c = 0$ whose roots are α and β.

Similarly the roots of the equation $by^2 + 2fy + c = 0$ are γ and δ. Hence, etc.

In the numerical part, $\alpha = 1$, $\beta = 2$, $\gamma = 1$, $\delta = 2$. The point (2, 2) is on the conic and this determines $h = 1/8$. Obtain the standard form and hence the eccentricity.

3. Using invariant theory, the standard form is $x^2 + 2y^2 = 1$. Hence, etc. (The result can also be obtained by a rotation through $-\theta$.)

4. The relative invariants are $I = 2(1 + \lambda^2)$; $J = (1 - \lambda^2)^2$; $\Delta = 2(1 + \lambda)^2 (1 - 2\lambda)$. Hence, etc.

5. Let the equation of the rectangular hyperbola referred to its axes be $2xy + k = 0$. The result follows by equating the absolute invariant Δ^2/J^3.

6. From (55.3) the area of the ellipse is $\pi\sqrt{-\Delta/\lambda_1 J}\sqrt{-\Delta/\lambda_2 J}$. Hence, etc.

Chapter X. Miscellaneous examples

1. Let the line pair be $l_1x + m_1y + n_1 = 0$ and $l_2x + m_2y + n_2 = 0$. Then $l_1l_2 = \rho a'$, etc., and the result follows from (69.7).

2. The pair of lines through the origin parallel to the tangents from A to the conic $S = 0$ is given by $S_A(ax^2 + 2hxy + by^2) - \{x(ax_A + hy_A + g) + y(hx_A + by_A + f)^2\} = 0$. The result follows from (16.1).

3. For a parabola $\mathscr{C} = 0$, and the equation of the isoptic locus is now quadratic. In virtue of (67.3) and $\mathscr{C} = 0$, the invariants of the isoptic locus are $\bar{I} = 4\Delta I(1 - \tan^2\alpha)$, $\bar{J} = -16I^2\Delta^2\tan^2\alpha$ and $\bar{\Delta} = 64\Delta^4\sec^2\alpha$. Hence, etc.

4. The centre of the isoptic locus of a parabola lies on the straight lines $\mathscr{G}(2\mathscr{G}x + 2\mathscr{F}y - \mathscr{A} - \mathscr{B})\tan^2\alpha + 2\Delta(ax + by + g) = 0$ and $\mathscr{F}(2\mathscr{G}x + 2\mathscr{F}y - \mathscr{A} - \mathscr{B})\tan^2\alpha + 2\Delta(hx + by + f) = 0$. Hence $\mathscr{F}(ax + hy + g) - \mathscr{G}(hx + by + f) = 0$, etc.

5. In this example $\mathscr{A} = \mathscr{B} = -1$; $\mathscr{C} = 1 - h^2$; $\mathscr{H} = h$; $\mathscr{F} = \mathscr{G} = 0$. Equations (76.3) then yield $(1 - h^2)(\alpha^2 - \beta^2) = 0$ and $(1 - h^2)\alpha\beta = -h$. Hence, etc. The values of the eccentricity follow immediately from (77.1).

6. The axis is obtained from (74.3), whilst (76.5) gives the focus. In this example $\mathscr{A} = \mathscr{C} = \mathscr{H} = 0$; $\mathscr{B} = -\rho^2$; $\mathscr{F} = -\lambda\mu\rho$; $\mathscr{G} = \rho\mu^2$.

7. By §73 the diameter conjugate to $lx + my + n = 0$ is $m(ax + hy + g) - l(hx + by + f) = 0$. Hence $lx + my + n = 0$ is identical with $l(ax + hy + g) + m(hx + by + f) = 0$. Compare coefficients, etc.

The principal axes are $\left(m\dfrac{\partial S}{\partial x} - l\dfrac{\partial S}{\partial y}\right)\left(l\dfrac{\partial S}{\partial x} + m\dfrac{\partial S}{\partial y}\right) = 0$. Further $a + hm/l = hl/m + b$. Hence, etc.

8. Let $y = gx$ be a normal through the origin. It intersects the conic at $P \equiv (-2fg/(a + 2hg + bg^2),\ -2fg^2/(a + 2hg + bg^2))$. The normal at P is $(hx_P + by_P + f)(x - x_P) - (ax_P + hy_P)(y - y_P) = 0$. Hence $h(x_P^2 - y_P^2) + (b - a)x_Py_P + fx_P = 0$. Substitute the coordinates of P and the required cubic is obtained.

9. Almost identical with Example 1.

10. By translating the origin, we have $x = at^2 + 2bt$, $y = a't^2 + 2b't$. Thus $(ay - a'x)^2 = 4(ab' - a'b)(b'x - by)$. The invariants are $I = a^2 + a'^2$ and $\Delta = -4(ab' - a'b)^4$ and so the semi-latus rectum is $\sqrt{-\Delta/I^3}$, etc.

11. The hyperbola conjugate to $S = 0$ is $S + k = 0$ since they have the same asymptotes. The standard forms are $x^2/a^2 - y^2/b^2 - 1 = 0$ and

$x^2/a^2 - y^2/b^2 + 1 = 0$. The invariants of the conjugate hyperbola are then I, J and $-\Delta$. Hence $-\Delta/IJ = (\Delta + kJ)/IJ$, etc.

Chapter XI

1. The polar of P with respect to $S + \lambda S' = 0$ is $T_P + \lambda T_P' = 0$ and passes through the fixed point $T_P = T_P' = 0$.

2. The centre of the conic is at $\partial S/\partial x + \lambda \partial S'/\partial x = 0$; $\partial S/\partial y + \lambda \partial S'/\partial y = 0$ and so the locus is the conic $\dfrac{\partial S}{\partial x}\dfrac{\partial S'}{\partial y} - \dfrac{\partial S'}{\partial x}\dfrac{\partial S}{\partial y} = 0$.

3. If $I = I' = 0$, then the corresponding invariant $I + \lambda I'$ of $S + \lambda S'$ is also zero.

4. The circle is $S + \lambda u^2 = 0$ if $a + \lambda l^2 = b + \lambda m^2$ and $h + \lambda lm = 0$. Hence $(a - b)/h = (l^2 - m^2)/lm$, etc.

5. The centre of the conic $S + \lambda T_A^2 = 0$ is at $\partial S/\partial x + \lambda T_A \partial S_A/\partial x_A = 0$ and $\partial S/\partial y + \lambda T_A \partial S_A/\partial y_A = 0$ and so the locus is the straight line $\dfrac{\partial S_A}{\partial y_A}\dfrac{\partial S}{\partial x} - \dfrac{\partial S_A}{\partial x_A}\dfrac{\partial S}{\partial y} = 0$.

Miscellaneous examples

1. The pencil of conics is $\lambda xy + (lx + my + n)^2 = 0$. By (71.1) the orthoptic locus is $(\lambda + 2lm)(x^2 + y^2) + 2n(mx + ly) = 0$, etc.

2. Let the vertices of the triangle be at $(\alpha_1, 0)$, $(\alpha_2, 0)$ and $(0, \beta)$. The required pencil is then $\lambda xy + (\beta x + \alpha_1 y - \alpha_1\beta)(\alpha_1 x - \beta y - \alpha_1\alpha_2) = 0$. The equations for the centre are $2\beta\alpha_1 x + y(\lambda + \alpha_1^2 - \beta^2) - \alpha_1\beta(\alpha_1 + \alpha_2) = 0$ and $-2\beta\alpha_1 y + x(\lambda_1 + \alpha_1^2 - \beta^2) + \alpha_1(\beta^2 - \alpha_1\alpha_2) = 0$. Eliminate λ and the locus is the circle $2\beta(x^2 + y^2) - \beta(\alpha_1 + \alpha_2)x + (\alpha_1\alpha_2 - \beta^2)y = 0$, etc.

4. The pencil of conics through P, Q, R and S is $S + \lambda T_A T_B = 0$. This conic passes through both A and B if $\lambda = -1/T_{AB}$.

5. Choose the straight line as x-axis and let its pole be at A. Then $y = 0$ and $T_A + \lambda T_A' = 0$ are the same line. Thus $ax_A + hy_A + g + \lambda(a'x_A + h'y_A + g') = 0$ and $gx_A + fy_A + c + \lambda(g'x_A + f'y_A + c') = 0$. Eliminate λ and the locus of A is clearly a conic.

6. (*a*) By Example 2, this page, the locus of centres is the conic $\dfrac{\partial S}{\partial x}\dfrac{\partial S'}{\partial y} - \dfrac{\partial S'}{\partial x}\dfrac{\partial S}{\partial y} = 0$. Its asymptotes are parallel to $(ah' - a'h)x^2 + (ab' - a'b)xy + (b'h - bh')y^2 = 0$. Hence its principal axes are parallel to the bisectors $(ab' - a'b)(x^2 - y^2) + 2\{h(a' + b') - h'(a + b)\}xy = 0$ (Example 4, page 19). The rectangular hyperbola of the pencil is $(a' + b')S - (a + b)S' = 0$. Hence, etc.

(*b*) Let A be the fixed point. Then $\dfrac{\partial S_A}{\partial x_A}\dfrac{\partial S_A'}{\partial y_A} = \dfrac{\partial S_A'}{\partial x_A}\dfrac{\partial S_A}{\partial y_A}$. The polar of A is $T_A + \lambda T_A' = 0$. Its gradient is $-\left(\dfrac{\partial S_A}{\partial x_A} + \lambda\dfrac{\partial S_A'}{\partial x_A}\right)\Big/\left(\dfrac{\partial S_A}{\partial y_A} + \lambda\dfrac{\partial S_A'}{\partial y_A}\right) = -\dfrac{\partial S_A'}{\partial x_A}\Big/\dfrac{\partial S_A'}{\partial y_A}$. Hence, etc.

7. Select the vertices of the parallelogram at $(\alpha, 0)$, $(-\alpha, 0)$, $(0, \beta)$ and $(0, -\beta)$. Then the pencil is $\lambda xy + \left(\dfrac{x}{\alpha} + \dfrac{y}{\beta} - 1\right)\left(\dfrac{x}{\alpha} + \dfrac{y}{\beta} + 1\right) = 0$. Hence, etc.

8. Choose the conics as $S \equiv ax^2 + by^2 + c = 0$ and $S' \equiv a'x^2 + b'y^2 + 2g'x + 2f'y + c' = 0$. Then the pencil $S + \lambda S' = 0$ contains a circle if $a + \lambda a' = b + \lambda b'$. Hence, etc.

9. Let the conics be $S_1 \equiv S + \lambda_1 u^2 = 0$ and $S_2 \equiv S + \lambda_2 u^2 = 0$, where $u = 0$ is the chord of contact. Let the two lines be $v = 0$ and $w = 0$. Then constants α, β, γ exist such that $\alpha u + \beta v + \gamma w = 0$. Consider $S_1 + kvw \equiv S_2 + (\lambda_1 - \lambda_2)u^2 - kv(\alpha u + \beta v)/\gamma$. Choose k so that $\gamma(\lambda_1 - \lambda_2)u^2 - k\alpha uv - k\beta v^2$ is a perfect square. That is, $k = -4\beta\gamma(\lambda_1 - \lambda_2)/\alpha^2$. Then $S_1 + kvw \equiv S_2 + \gamma(\lambda_1 - \lambda_2)(u + 2\beta v/\alpha)^2 = 0$ is a conic having double contact with $S_2 = 0$ as required.

Chapter XII. Miscellaneous examples

1. Let the roots be t_1, t_1' and t_2, t_2' respectively. Then $t_1 + t_1' = -2h_1/a_1$, etc. Now $\{t_1, t_1';\ t_2, t_2'\} = (t_1 - t_2)(t_1' - t_2')/(t_1 - t_2')(t_1' - t_2) = [t_1t_1' + t_2t_2' - (t_1t_2' + t_2t_1')]/[t_1t_1' + t_2t_2' - (t_1t_2 + t_1't_2')]$. But $t_1t_2 + t_1't_2' + t_1t_2' + t_2t_1' = (t_1 + t_1')(t_2 + t_2')$ and $t_1t_2 + t_1't_2' - t_1t_2' - t_2t_1' = (t_1 - t_1')(t_2 - t_2')$. Further, $(t_1 - t_1')^2 = (t_1 + t_1')^2 - 4t_1t_1'$. Hence, etc.

2. If γ corresponds to δ, the result is a direct consequence of $\{x, \gamma;\ \alpha, \beta\} = \{y, \delta;\ \alpha, \beta\}$.

Alternatively: α and β are the roots of $at^2 + (b + c)t + d = 0$. Thus $\alpha + \beta = -(b + c)/a$ and $\alpha\beta = d/a$. Write $\lambda = c/a$, then $b/a = -(\lambda + \alpha + \beta)$ and correspondence becomes $xy - (\lambda + \alpha + \beta)x + \lambda y + \alpha\beta = 0$. That is $(x - \alpha)(y - \beta) = (\alpha + \lambda)(x - y)$. Similarly $(x - \beta)(y - \alpha) = (\beta + \lambda)(x - y)$. Hence, by division, etc.

3. If the double points coincide at α, then by Example 2, $(x - \alpha)(y - \alpha) = (\alpha + \lambda)(x - y)$. Thus $1/(x - \alpha) - 1/(y - \alpha) = -1/(\alpha + \lambda) =$ constant, as required.

4. Let the roots be t_1 and t_2. Then $t_1 + t_2 = -2(h_1 + \lambda h_2)/(a_1 + \lambda a_2)$ and $t_1t_2 = (b_1 + \lambda b_2)/(a_1 + \lambda a_2)$. Elimination of λ yields the involution $2(a_1h_2 - a_2h_1)t_1t_2 + (a_1b_2 - a_2b_1)(t_1 + t_2) + 2(b_2h_1 - b_1h_2) = 0$.

5. If the roots of $Q_1 = 0$, $Q_2 = 0$ and $Q_3 = 0$ are in involution, then by Example 4 $\mu Q_3 \equiv Q_1 + \lambda Q_2$. Equate coefficients and eliminate μ and λ to obtain the determinantal result given in the answers.

6. Let the double points be at ξ. Then the roots of $(t - \xi)^2 = 0$ and $Q_1 = 0$, $Q_2 = 0$ are in involution. Thus the double points by Example 5 are the roots of the determinantal equation given in the answers.

7. The gradients of the normals are ∞ and the three roots of the cubic $2hg^3 + (2a - b)g^2 + a = 0$ (Example 8, page 90). The four normals then form a harmonic pencil if $\{g_1, g_2;\ g_3, \infty\} = -1$. That is $3g_3 = g_1 + g_2 + g_3 = -(2a - b)/2h$. Hence, etc.

8. The chords $2x - (t_1 + t_2)y + 2at_1t_2 = 0$ and $2x - (t_3 + t_4)y + 2at_3t_4 = 0$ are conjugate with respect to $y^2 = 4ax$ if $2(t_1t_2 + t_3t_4) = (t_1 + t_2)(t_3 + t_4)$. That is, $\{t_1, t_2;\ t_3, t_4\} = -1$ as required.

9. As Example 8 with chords $x + t_1t_2y - c(t_1 + t_2) = 0$, etc.

Chapter XIII

9. The conic $\Sigma + \lambda\Sigma' = 0$ is a rectangular hyperbola if $(\mathscr{B} + \lambda\mathscr{B}')(\mathscr{C} + \lambda\mathscr{C}') - (\mathscr{F} + \lambda\mathscr{F}')^2 + (\mathscr{C} + \lambda\mathscr{C}')(\mathscr{A} + \lambda\mathscr{A}') - (\mathscr{G} + \lambda\mathscr{G}')^2 = 0$. Hence, etc.

The conic is a parabola if $\mathscr{C} + \lambda\mathscr{C}' = 0$. Hence, etc.

10. The centre is the pole of the line at infinity (0, 0, 1). The required result follows from (92.1).

The locus of centres is the line $(\mathscr{F}\mathscr{C}' - \mathscr{F}'\mathscr{C}, \mathscr{C}\mathscr{G}' - \mathscr{C}'\mathscr{G}, \mathscr{G}\mathscr{F}' - \mathscr{G}'\mathscr{F})$. The opposite vertices are degenerate line-conics of the range, and their mid-points correspond to their centres. Hence, etc.

14. The pole of $lx + my + n = 0$ with respect to the conic (98.2) is at $(-l(a^2 + \lambda)/n, -m(b^2 + \lambda)/n)$. Thus the locus of poles is the straight line $x/l - y/m = (b^2 - a^2)/n$, which is orthogonal to $lx + my + n = 0$. There is a conic of the confocal system touching $lx + my + n = 0$ and its pole is the point of contact. Hence, etc.

15. For the confocals λ_1 and λ_2, the orthogonal tangents (see Example 2, on page 36) are $y = gx + \sqrt{(a^2 + \lambda_1)g^2 + b^2 + \lambda_1}$ and $y = -x/g + \sqrt{(a^2 + \lambda_2)/g^2 + b^2 + \lambda_2}$. Elimination of g yields the required result $x^2 + y^2 = a^2 + b^2 + \lambda_1 + \lambda_2$.

16. The pole of $lx + my + n = 0$ with respect to the parabola (99.1) is at $(n/l - 2\mu, -2m\mu/l)$. Thus the locus of poles is $x/l - y/m = n/l^2$. Hence, etc.

17. The orthogonal tangents to the confocals μ_1 and μ_2 are $y = gx + \mu_1 g + \mu_1/g$ and $y - x/g - \mu_2 g - \mu_2/g$. Hence the locus is $x + \mu_1 + \mu_2 = 0$ as required.

Miscellaneous examples

1. The line-equation of $ax^2 + 2hxy + by^2 = 1$ is $bl^2 + am^2 - (ab - h^2)n^2 - 2hlm = 0$. Hence the confocals are $(b + k)l^2 + (a + k)m^2 - (ab - h^2)n^2 - 2hlm = 0$. Now $\nu l^2 + \lambda m^2 - (\lambda\nu - \mu^2)n^2 - 2\mu lm = 0$ is a member of the confocal system. Hence $(b + k)/\nu = (a + k)/\lambda = (ab - h^2)/(\lambda\nu - \mu^2) = h/\mu$, etc.

Further, the point-equation of the confocal system is $(ab - h^2)[(a + k)x^2 + 2hxy + (b + k)y^2] - \{(a + k)(b + k) - h^2\} = 0$; and so its invariants are $\bar{I} = J(2k + I)$; $\bar{J} = J^2(k^2 + kI + J)$ and $\bar{\Delta} = -J^2(k^2 + kI + J)^2$. Equate the absolute invariants with those of $\lambda x^2 + 2\mu xy + \nu y^2 = 1$ and eliminate k to obtain the required result.

2. The conditions are $a = b$ and $h = 0$. In virtue of (67.3) the required result is obtained.

3. The corresponding line-equations are $pl^2 + n^2 + 2\alpha ln + 2\beta mn = 0$ and $qm^2 + n^2 + 2\alpha ln + 2\beta mn = 0$. Hence, etc.

5. The line-equation of the conics are $bcl^2 + (ca - g^2)m^2 + abn^2 - 2bgln = 0$ and $(bc - f^2)l^2 + cam^2 + abn^2 - 2afmn = 0$. The parabola of the range is $f^2l^2 - g^2m^2 - 2bgln + 2afmn = 0$. Its directrix is obtained from (71.2).

7. The orthoptic circle of the conic $\Sigma + \lambda\Sigma' = 0$ is $(\mathscr{C} + \lambda\mathscr{C}')(x^2 + y^2) - 2(\mathscr{G} + \lambda\mathscr{G}')x - 2(\mathscr{F} + \lambda\mathscr{F}')y + \mathscr{A} + \lambda\mathscr{A}' + \mathscr{B} + \lambda\mathscr{B}' = 0$. The radical axis corresponds to $\lambda = -\mathscr{C}/\mathscr{C}'$. This value of λ gives the parabola of the range. Hence, etc.

8. The reader may tackle this problem separately for central conics and parabolas.

Alternatively, the polar of A with respect to the confocal (97.1) has line-coordinates $l = \Delta(ax_A + hy_A + g) + \lambda(\mathscr{C}x_A - \mathscr{G})$; $m = \Delta(hx_A + by_A + f) + \lambda(\mathscr{C}y_A - \mathscr{G})$ and $n = \Delta(gx_A + fy_A + c) + \lambda(\mathscr{A} + \mathscr{B} - \mathscr{G}x_A - \mathscr{F}y_A) + \lambda^2$. Elimination of λ from the homogeneous equations $l/m = \ldots$ and $n/m = \ldots$ yields a quadratic in l, m, n which has no term in n^2 and so the envelope is a parabola.

Chapter XIV

6. Choose ABC as triangle of reference. The tangent at A to the conic (110.2) is $g\zeta + h\eta = 0$. Thus $D \equiv (0, g, -h)$. Similarly $E \equiv (f, 0, -h)$ and $F \equiv (f, -g, 0)$. Hence, etc.

7. The points of contact are $(0, \gamma, \beta)$, $(\gamma, 0, \alpha)$ and $(\beta, \alpha, 0)$ and so the straight lines in the question are $\beta\eta - \gamma\zeta = 0$; $\gamma\zeta - \alpha\xi = 0$ and $\alpha\xi - \beta\eta = 0$. Hence, etc.

8. Select one of the triangles, say ABC as triangle of reference. Call the other triangle PQR. Since PQR is a self-conjugate triangle with respect to the conic (110.5), it follows that $a\xi_Q\xi_R + b\eta_Q\eta_R + c\zeta_Q\zeta_R = 0$, etc. Hence

$$\begin{vmatrix} \xi_Q\xi_R & \eta_Q\eta_R & \zeta_Q\zeta_R \\ \xi_R\xi_P & \eta_R\eta_P & \zeta_R\zeta_P \\ \xi_P\xi_Q & \eta_P\eta_Q & \zeta_P\zeta_Q \end{vmatrix} = 0. \text{ Thus } \begin{vmatrix} \frac{1}{\xi_P} & \frac{1}{\eta_P} & \frac{1}{\zeta_P} \\ \frac{1}{\xi_Q} & \frac{1}{\eta_Q} & \frac{1}{\zeta_Q} \\ \frac{1}{\xi_R} & \frac{1}{\eta_R} & \frac{1}{\zeta_R} \end{vmatrix} = 0.$$

$$\text{That is, } \begin{vmatrix} \eta_P\zeta_P & \zeta_P\xi_P & \xi_P\eta_P \\ \eta_Q\zeta_Q & \zeta_Q\xi_Q & \xi_Q\eta_Q \\ \eta_R\zeta_R & \zeta_R\xi_R & \xi_R\eta_R \end{vmatrix} = 0.$$

Thus there exists a conic $f\eta\zeta + g\eta\zeta + h\xi\eta = 0$ which passes through P, Q, R, A, B and C.

10. Let the involution be $at_1t_2 + b(t_1 + t_2) + d = 0$. Then the chord (Example 9), $x - (t_1 + t_2)y + t_1t_2z = 0$, passes through the fixed point $(d, -b, a)$ as required.

14. The pole of $\lambda\xi + \mu\eta + \nu\zeta = 0$ with respect to the conic (111.3) is at $(\lambda\rho(1 + \rho), \mu(1 + \rho), -\nu\rho)$ and so the required locus is $\lambda\eta\zeta + \mu\zeta\xi + \nu\xi\eta = 0$. Hence, etc.

15. The pole (α, β, γ) of $\lambda\xi + \mu\eta + \nu\zeta = 0$ with respect to the conic (111.5) satisfies $(\beta + \gamma)/\lambda = (\rho\gamma + \alpha)/\mu = (\rho\beta + \alpha)/\nu$. Thus the required locus is $\mu\eta^2 - \nu\zeta^2 + (\mu - \nu)\eta\zeta + \lambda\xi(\zeta - \eta) = 0$. Hence, etc.

16. The pole of $\lambda\xi + \mu\eta + \nu\zeta = 0$ with respect to the conic (111.7) is at $(\rho\nu, \mu, \rho\lambda)$ and so the required locus is $\lambda\xi - \nu\zeta = 0$. Hence, etc.

17. The pole (α, β, γ) of $\lambda\xi + \mu\eta + \nu\zeta = 0$ with respect to the conic (111.9) satisfies $h\eta/\lambda = (h\xi + \rho\zeta)/\mu = (\rho\eta + c\zeta)/\nu$ and so the required locus is $\mu h\eta^2 + \lambda c\zeta^2 - \nu h\eta\zeta - h\lambda\xi\eta = 0$. Hence, etc.

18. The pole (α, β, γ) of $\lambda\xi + \mu\eta + \nu\zeta = 0$ with respect to the conic (111.10) satisfies $h\beta/\lambda = (h\alpha + \rho\beta + f\gamma)/\mu = (f\beta + c\gamma)/\nu$, and so the required locus is $(\nu h - \lambda f)\eta - \lambda c\zeta = 0$. Hence, etc.

Miscellaneous examples

1. Select the diagonal point triangle as the triangle of reference. Then two pairs of opposite vertices are $(1, 1, 1)$, $(-1, 1, 1)$ and $(1, -1, 1)$, $(1, 1, -1)$, whilst the third pair is $(0, 1, 0)$ and $(0, 0, 1)$. The conjugacy conditions for $S = 0$ are $-a + b + c + 2f = 0$ and $a - b - c + 2f = 0$. Thus $f = 0$. Hence, etc.

2. Let the conic be (111.4). Choose the unit point so that the fixed straight lines is $\eta = \zeta$. Then $P \equiv (\rho, -(g+h), -(g+h))$, and so the tangent at P is $(g+h)\xi + \rho g\eta + \rho h\zeta = 0$. Thus $Q \equiv (\rho gh, -h(g+h)^2, -g(g+h)^2)$ Now obtain the equation of PQ and verify that $(0, h^2, g^2)$ lies on it, as required.

3. Select ABC as the triangle of reference and the inscribed conic has the equation $\alpha^2\xi^2 + \beta^2\eta^2 + \gamma^2\zeta^2 - 2\beta\gamma\eta\zeta - 2\gamma\alpha\zeta\xi - 2\alpha\beta\xi\eta = 0$. Then $D \equiv (0, \gamma, \beta)$ and so $D' \equiv (0, \gamma, -\beta)$. The pair of tangents from D' are $\xi = 0$ and $\alpha\xi - 2\beta\eta - 2\gamma\zeta = 0$. Thus $P \equiv (2\beta\gamma, \gamma\alpha, \alpha\beta)$. Further, $X \equiv (4\beta\gamma, \gamma\alpha, \alpha\beta)$. Hence, etc.

Similarly $Q \equiv (\beta\gamma, 2\gamma\alpha, \alpha\beta)$ and $R \equiv (\beta\gamma, \gamma\alpha, 2\alpha\beta)$. Let the locus of R be the fixed line $a\xi + b\eta + c\zeta = 0$. Therefore, $a\beta\gamma + b\gamma\alpha + 2c\alpha\beta = 0$. The line PQ has line-coordinates $(-\alpha^2\beta\gamma, -\alpha\beta\gamma^2, 3\alpha\beta\gamma^2)$. That is $(-\alpha, -\beta, 3\gamma)$. Hence the line equation of its envelope is $a\mu\nu + b\nu\lambda - 6c\lambda\mu = 0$. Hence, etc.

4. From Example 9, page 127, the chord joining t_1 and t_2 is $\xi - (t_1 + t_2)\eta + t_1t_2\zeta = 0$. Let the homographic correspondence be $at_1t_2 + bt_1 + ct_2 + d = 0$. The line-coordinates of the chord are $\lambda = 1$, $\mu = -(t_1 + t_2)$, $\nu = t_1t_2$. Thus $\nu/\lambda = t_1t_2$ and $\mu/\lambda = -(t_1 + t_2)$. The elimination of t_1 and t_2 yields $(b-c)^2\nu\lambda + (b\mu - a\nu - d\lambda)(c\mu - a\nu - d\lambda) = 0$, which can be written $(b-c)^2(\mu^2 - 4\nu\lambda) - [2d\lambda - (b+c)\mu + 2a\nu]^2 = 0$. This then represents a conic having double contact with $\mu^2 - 4\nu\lambda = 0$. In point-coordinates this conic is $\eta^2 - \xi\zeta = 0$. Hence, etc.

5. Let the conic be (111.7). The pair of tangents from C is $\eta = 0$ and $2ch\xi - \rho^2\eta = 0$. Hence $D \equiv (\rho^2, 2ch, -2h\rho)$ and so its locus is $2h\xi\eta - c\zeta^2 = 0$, as required.

Further, $E \equiv (0, c, -2\rho)$. Thus the line-coordinates of DE are $(2ch, -2\rho^2, -c\rho)$ and so its envelope is $c\lambda\mu + 4h\nu^2 = 0$, as required.

6. Let the pencil be (111.8). Then $S_1 \equiv (f\eta + c\zeta)^2 + 2hc\xi\eta = 0$ and $S_2 \equiv c\zeta^2 + 2f\eta\zeta + 2h\xi\eta = 0$. Thus $L \equiv (0, c, -f)$ and $M \equiv (0, c, -2f)$. Hence, etc.

7. From Fig. 61, with ABC as the triangle of reference, $P \equiv (1, 1, 1)$, $Q \equiv (1, 1, -1)$, $R \equiv (-1, 1, 1)$, $S \equiv (1, -1, 1)$. Let $\Sigma \equiv \xi^2 + \rho\eta^2 - (1+\rho)\zeta^2 = 0$. Then $\Sigma_1 \equiv \xi^2 + \rho\eta^2 - (1+\rho)\zeta^2 + k(\xi - \eta)^2 = 0$. Since A lies on Σ_1 it follows that $k = -1$ and so $\Sigma_1 \equiv (\rho - 1)\eta^2 - (\rho + 1)\zeta^2 + 2\xi\eta = 0$. Similarly $\Sigma_2 \equiv (\rho - 1)\eta^2 - (\rho + 1)\zeta^2 - 2\xi\eta = 0$. The common tangent at A is $\eta = 0$ as required for single contact at A. In line-coordinates Σ_1 and Σ_2 become $(1 - \rho^2)\lambda^2 - \nu^2 + 2(\rho + 1)\mu\nu = 0$ and $(1 - \rho^2)\lambda^2 - \nu^2 - 2(\rho + 1)\mu\nu = 0$ and so the common tangents not through A are $\xi \pm \sqrt{1 - \rho^2}\,\zeta = 0$, as required.

8. Let the vertices taken in order be $ANBLCM$. With ABC as triangle of reference, let the circumscribed conic be (110.1). Then $f\eta_L\zeta_L + g\zeta_L\xi_L + h\xi_L\eta_L = 0$ and two similar equations in the indices M and N hold. Eliminate $f:g:h$ and the result is

$$\begin{vmatrix} \eta_L\zeta_L & \zeta_L\xi_L & \xi_L\eta_L \\ \eta_M\zeta_M & \zeta_M\xi_M & \xi_M\eta_M \\ \eta_N\zeta_N & \zeta_N\xi_N & \xi_N\eta_N \end{vmatrix} = 0.$$

The straight lines CM and BN intersect at $(\xi_M\xi_N, \eta_M\xi_N, \xi_M\zeta_N)$. The other two intersections are $(\eta_N\xi_L, \eta_N\eta_L, \zeta_N\eta_L)$ and $(\xi_L\zeta_M, \zeta_L\eta_M, \zeta_L\zeta_M)$. These three points are collinear if

$$\begin{vmatrix} \xi_M\xi_N & \eta_M\zeta_N & \xi_M\zeta_N \\ \eta_N\xi_L & \eta_N\eta_L & \zeta_N\eta_L \\ \xi_L\zeta_M & \zeta_L\eta_M & \zeta_L\zeta_M \end{vmatrix} = 0.$$

Divide corresponding rows of this determinant by $\xi_M\xi_N$, $\eta_N\eta_L$ and $\zeta_L\zeta_M$ respectively. Then multiply columns by $\eta_L\zeta_L$, $\zeta_M\xi_M$ and $\xi_N\eta_N$ respectively and the two determinantal equations are seen to be equivalent. Hence, etc.

9. Select the triangle as triangle of reference and choose the unit point so that the circumscribed conic S is $\eta\zeta + \zeta\xi + \xi\eta = 0$. Let the *line-equation* of the inscribed conic Σ be $\alpha\mu\nu + \beta\nu\lambda + \gamma\lambda\mu = 0$. We may take $(\lambda + 1, \lambda^2 + \lambda, -\lambda)$ as parametric coordinates of a point on S. Let P, Q, R correspond to λ_1, λ_2, λ_3. Then the equation of QR is $\lambda_2\lambda_3\xi + \eta + (\lambda_2 + 1)(\lambda_3 + 1)\zeta = 0$. This line is tangent to Σ if $u \equiv \alpha(\lambda_2 + 1)(\lambda_3 + 1) + \beta\lambda_2\lambda_3(\lambda_2 + 1)(\lambda_3 + 1) + \gamma\lambda_2\lambda_3 = 0$. The lines RP and PQ are tangent to Σ if the corresponding expressions v and w obtained by cyclic interchange are zero. Verify that $(\lambda_1^2 + \lambda_1)(\lambda_2 - \lambda_3)u + (\lambda_2^2 + \lambda_2)(\lambda_3 - \lambda_1)v + (\lambda_3^2 + \lambda_3)(\lambda_1 - \lambda_2)w = 0$. Thus $w = 0$ holds if both $u = 0$ and $v = 0$. Hence, etc.

INDEX

Math–Geometry and Topology

ELEMENTARY CONCEPTS OF TOPOLOGY, Paul Alexandroff. Elegant, intuitive approach to topology from set-theoretic topology to Betti groups; how concepts of topology are useful in math and physics. 25 figures. 57pp. 5⅜ x 8½. 0-486-60747-X

COMBINATORIAL TOPOLOGY, P. S. Alexandrov. Clearly written, well-organized, three-part text begins by dealing with certain classic problems without using the formal techniques of homology theory and advances to the central concept, the Betti groups. Numerous detailed examples. 654pp. 5⅜ x 8½. 0-486-40179-0

EXPERIMENTS IN TOPOLOGY, Stephen Barr. Classic, lively explanation of one of the byways of mathematics. Klein bottles, Moebius strips, projective planes, map coloring, problem of the Koenigsberg bridges, much more, described with clarity and wit. 43 figures. 210pp. 5⅜ x 8½. 0-486-25933-1

THE GEOMETRY OF RENÉ DESCARTES, René Descartes. The great work founded analytical geometry. Original French text, Descartes's own diagrams, together with definitive Smith-Latham translation. 244pp. 5⅜ x 8½. 0-486-60068-8

EUCLIDEAN GEOMETRY AND TRANSFORMATIONS, Clayton W. Dodge. This introduction to Euclidean geometry emphasizes transformations, particularly isometries and similarities. Suitable for undergraduate courses, it includes numerous examples, many with detailed answers. 1972 ed. viii+296pp. 6⅛ x 9¼. 0-486-43476-1

PRACTICAL CONIC SECTIONS: THE GEOMETRIC PROPERTIES OF ELLIPSES, PARABOLAS AND HYPERBOLAS, J. W. Downs. This text shows how to create ellipses, parabolas, and hyperbolas. It also presents historical background on their ancient origins and describes the reflective properties and roles of curves in design applications. 1993 ed. 98 figures. xii+100pp. 6½ x 9¼. 0-486-42876-1

THE THIRTEEN BOOKS OF EUCLID'S ELEMENTS, translated with introduction and commentary by Sir Thomas L. Heath. Definitive edition. Textual and linguistic notes, mathematical analysis. 2,500 years of critical commentary. Unabridged. 1,414pp. 5⅜ x 8½.. Three-vol. set.
Vol. I: 0-486-60088-2 Vol. II: 0-486-60089-0 Vol. III: 0-486-60090-4

SPACE AND GEOMETRY: IN THE LIGHT OF PHYSIOLOGICAL, PSYCHOLOGICAL AND PHYSICAL INQUIRY, Ernst Mach. Three essays by an eminent philosopher and scientist explore the nature, origin, and development of our concepts of space, with a distinctness and precision suitable for undergraduate students and other readers. 1906 ed. vi+148pp. 5⅜ x 8½. 0-486-43909-7

GEOMETRY OF COMPLEX NUMBERS, Hans Schwerdtfeger. Illuminating, widely praised book on analytic geometry of circles, the Moebius transformation, and two-dimensional non-Euclidean geometries. 200pp. 5⅝ x 8¼. 0-486-63830-8

DIFFERENTIAL GEOMETRY, Heinrich W. Guggenheimer. Local differential geometry as an application of advanced calculus and linear algebra. Curvature, transformation groups, surfaces, more. Exercises. 62 figures. 378pp. 5⅜ x 8½. 0-486-63433-7

History of Math

THE WORKS OF ARCHIMEDES, Archimedes (T. L. Heath, ed.). Topics include the famous problems of the ratio of the areas of a cylinder and an inscribed sphere; the measurement of a circle; the properties of conoids, spheroids, and spirals; and the quadrature of the parabola. Informative introduction. clxxxvi+326pp. 5⅜ x 8½. 0-486-42084-1

A SHORT ACCOUNT OF THE HISTORY OF MATHEMATICS, W. W. Rouse Ball. One of clearest, most authoritative surveys from the Egyptians and Phoenicians through 19th-century figures such as Grassman, Galois, Riemann. Fourth edition. 522pp. 5⅜ x 8½. 0-486-20630-0

THE HISTORY OF THE CALCULUS AND ITS CONCEPTUAL DEVELOPMENT, Carl B. Boyer. Origins in antiquity, medieval contributions, work of Newton, Leibniz, rigorous formulation. Treatment is verbal. 346pp. 5⅜ x 8½. 0-486-60509-4

THE HISTORICAL ROOTS OF ELEMENTARY MATHEMATICS, Lucas N. H. Bunt, Phillip S. Jones, and Jack D. Bedient. Fundamental underpinnings of modern arithmetic, algebra, geometry and number systems derived from ancient civilizations. 320pp. 5⅜ x 8½. 0-486-25563-8

A HISTORY OF MATHEMATICAL NOTATIONS, Florian Cajori. This classic study notes the first appearance of a mathematical symbol and its origin, the competition it encountered, its spread among writers in different countries, its rise to popularity, its eventual decline or ultimate survival. Original 1929 two-volume edition presented here in one volume. xxviii+820pp. 5⅜ x 8½. 0-486-67766-4

GAMES, GODS & GAMBLING: A HISTORY OF PROBABILITY AND STATISTICAL IDEAS, F. N. David. Episodes from the lives of Galileo, Fermat, Pascal, and others illustrate this fascinating account of the roots of mathematics. Features thought-provoking references to classics, archaeology, biography, poetry. 1962 edition. 304pp. 5⅜ x 8½. (Available in U.S. only.) 0-486-40023-9

OF MEN AND NUMBERS: THE STORY OF THE GREAT MATHEMATICIANS, Jane Muir. Fascinating accounts of the lives and accomplishments of history's greatest mathematical minds–Pythagoras, Descartes, Euler, Pascal, Cantor, many more. Anecdotal, illuminating. 30 diagrams. Bibliography. 256pp. 5⅜ x 8½. 0-486-28973-7

HISTORY OF MATHEMATICS, David E. Smith. Nontechnical survey from ancient Greece and Orient to late 19th century; evolution of arithmetic, geometry, trigonometry, calculating devices, algebra, the calculus. 362 illustrations. 1,355pp. 5⅜ x 8½. Two-vol. set. Vol. I: 0-486-20429-4 Vol. II: 0-486-20430-8

A CONCISE HISTORY OF MATHEMATICS, Dirk J. Struik. The best brief history of mathematics. Stresses origins and covers every major figure from ancient Near East to 19th century. 41 illustrations. 195pp. 5⅜ x 8½. 0-486-60255-9

Mathematics

FUNCTIONAL ANALYSIS (Second Corrected Edition), George Bachman and Lawrence Narici. Excellent treatment of subject geared toward students with background in linear algebra, advanced calculus, physics and engineering. Text covers introduction to inner-product spaces, normed, metric spaces, and topological spaces; complete orthonormal sets, the Hahn-Banach Theorem and its consequences, and many other related subjects. 1966 ed. 544pp. 6⅛ x 9¼. 0-486-40251-7

ASYMPTOTIC EXPANSIONS OF INTEGRALS, Norman Bleistein & Richard A. Handelsman. Best introduction to important field with applications in a variety of scientific disciplines. New preface. Problems. Diagrams. Tables. Bibliography. Index. 448pp. 5⅜ x 8½. 0-486-65082-0

VECTOR AND TENSOR ANALYSIS WITH APPLICATIONS, A. I. Borisenko and I. E. Tarapov. Concise introduction. Worked-out problems, solutions, exercises. 257pp. 5⅝ x 8¼. 0-486-63833-2

AN INTRODUCTION TO ORDINARY DIFFERENTIAL EQUATIONS, Earl A. Coddington. A thorough and systematic first course in elementary differential equations for undergraduates in mathematics and science, with many exercises and problems (with answers). Index. 304pp. 5⅜ x 8½. 0-486-65942-9

FOURIER SERIES AND ORTHOGONAL FUNCTIONS, Harry F. Davis. An incisive text combining theory and practical example to introduce Fourier series, orthogonal functions and applications of the Fourier method to boundary-value problems. 570 exercises. Answers and notes. 416pp. 5⅜ x 8½. 0-486-65973-9

COMPUTABILITY AND UNSOLVABILITY, Martin Davis. Classic graduate-level introduction to theory of computability, usually referred to as theory of recurrent functions. New preface and appendix. 288pp. 5⅜ x 8½. 0-486-61471-9

ASYMPTOTIC METHODS IN ANALYSIS, N. G. de Bruijn. An inexpensive, comprehensive guide to asymptotic methods–the pioneering work that teaches by explaining worked examples in detail. Index. 224pp. 5⅜ x 8½ 0-486-64221-6

APPLIED COMPLEX VARIABLES, John W. Dettman. Step-by-step coverage of fundamentals of analytic function theory–plus lucid exposition of five important applications: Potential Theory; Ordinary Differential Equations; Fourier Transforms; Laplace Transforms; Asymptotic Expansions. 66 figures. Exercises at chapter ends. 512pp. 5⅜ x 8½. 0-486-64670-X

INTRODUCTION TO LINEAR ALGEBRA AND DIFFERENTIAL EQUATIONS, John W. Dettman. Excellent text covers complex numbers, determinants, orthonormal bases, Laplace transforms, much more. Exercises with solutions. Undergraduate level. 416pp. 5⅜ x 8½. 0-486-65191-6

RIEMANN'S ZETA FUNCTION, H. M. Edwards. Superb, high-level study of landmark 1859 publication entitled "On the Number of Primes Less Than a Given Magnitude" traces developments in mathematical theory that it inspired. xiv+315pp. 5⅜ x 8½. 0-486-41740-9

TENSOR CALCULUS, J.L. Synge and A. Schild. Widely used introductory text covers spaces and tensors, basic operations in Riemannian space, non-Riemannian spaces, etc. 324pp. 5⅜ x 8¼. 0-486-63612-7

ORDINARY DIFFERENTIAL EQUATIONS, Morris Tenenbaum and Harry Pollard. Exhaustive survey of ordinary differential equations for undergraduates in mathematics, engineering, science. Thorough analysis of theorems. Diagrams. Bibliography. Index. 818pp. 5⅜ x 8½. 0-486-64940-7

INTEGRAL EQUATIONS, F. G. Tricomi. Authoritative, well-written treatment of extremely useful mathematical tool with wide applications. Volterra Equations, Fredholm Equations, much more. Advanced undergraduate to graduate level. Exercises. Bibliography. 238pp. 5⅜ x 8½. 0-486-64828-1

FOURIER SERIES, Georgi P. Tolstov. Translated by Richard A. Silverman. A valuable addition to the literature on the subject, moving clearly from subject to subject and theorem to theorem. 107 problems, answers. 336pp. 5⅜ x 8½. 0-486-63317-9

INTRODUCTION TO MATHEMATICAL THINKING, Friedrich Waismann. Examinations of arithmetic, geometry, and theory of integers; rational and natural numbers; complete induction; limit and point of accumulation; remarkable curves; complex and hypercomplex numbers, more. 1959 ed. 27 figures. xii+260pp. 5⅜ x 8½. 0-486-63317-9

POPULAR LECTURES ON MATHEMATICAL LOGIC, Hao Wang. Noted logician's lucid treatment of historical developments, set theory, model theory, recursion theory and constructivism, proof theory, more. 3 appendixes. Bibliography. 1981 edition. ix + 283pp. 5⅜ x 8½. 0-486-67632-3

CALCULUS OF VARIATIONS, Robert Weinstock. Basic introduction covering isoperimetric problems, theory of elasticity, quantum mechanics, electrostatics, etc. Exercises throughout. 326pp. 5⅜ x 8½. 0-486-63069-2

THE CONTINUUM: A CRITICAL EXAMINATION OF THE FOUNDATION OF ANALYSIS, Hermann Weyl. Classic of 20th-century foundational research deals with the conceptual problem posed by the continuum. 156pp. 5⅜ x 8½. 0-486-67982-9

CHALLENGING MATHEMATICAL PROBLEMS WITH ELEMENTARY SOLUTIONS, A. M. Yaglom and I. M. Yaglom. Over 170 challenging problems on probability theory, combinatorial analysis, points and lines, topology, convex polygons, many other topics. Solutions. Total of 445pp. 5⅜ x 8½. Two-vol. set.
Vol. I: 0-486-65536-9 Vol. II: 0-486-65537-7